Pitman Research Notes in Mathematics Series

Submission of proposals for consideration
Suggestions for publication, in the form of outlines and representative samples, are invited by the Editorial Board for assessment. Intending authors should approach one of the main editors or another member of the Editorial Board, citing the relevant AMS subject classifications. Alternatively, outlines may be sent directly to the publisher's offices. Refereeing is by members of the board and other mathematical authorities in the topic concerned, throughout the world.

Preparation of accepted manuscripts
On acceptance of a proposal, the publisher will supply full instructions for the preparation of manuscripts in a form suitable for direct photo-lithographic reproduction. Specially printed grid sheets are provided and a contribution is offered by the publisher towards the cost of typing. Word processor output, subject to the publisher's approval, is also acceptable.

Illustrations should be prepared by the authors, ready for direct reproduction without further improvement. The use of hand-drawn symbols should be avoided wherever possible, in order to maintain maximum clarity of the text.

The publisher will be pleased to give any guidance necessary during the preparation of a typescript, and will be happy to answer any queries.

Important note
In order to avoid later retyping, intending authors are strongly urged not to begin final preparation of a typescript before receiving the publisher's guidelines and special paper. In this way it is hoped to preserve the uniform appearance of the series.

Longman Scientific & Technical
Longman House
Burnt Mill
Harlow, Essex, UK
(tel (0279) 26721)

Longman Scientific & Technical
Churchill Livingstone Inc.
1560 Broadway
New York, NY 10036, USA
(tel (212) 819-5453)

Titles in this series

1 Improperly posed boundary value problems
A Carasso and A P Stone

2 Lie algebras generated by finite dimensional ideals
I N Stewart

3 Bifurcation problems in nonlinear elasticity
R W Dickey

4 Partial differential equations in the complex domain
D L Colton

5 Quasilinear hyperbolic systems and waves
A Jeffrey

6 Solution of boundary value problems by the method of integral operators
D L Colton

7 Taylor expansions and catastrophes
T Poston and I N Stewart

8 Function theoretic methods in differential equations
R P Gilbert and R J Weinacht

9 Differential topology with a view to applications
D R J Chillingworth

10 Characteristic classes of foliations
H V Pittie

11 Stochastic integration and generalized martingales
A U Kussmaul

12 Zeta-functions: An introduction to algebraic geometry
A D Thomas

13 Explicit *a priori* inequalities with applications to boundary value problems
V G Sigillito

14 Nonlinear diffusion
W E Fitzgibbon III and H F Walker

15 Unsolved problems concerning lattice points
J Hammer

16 Edge-colourings of graphs
S Fiorini and R J Wilson

17 Nonlinear analysis and mechanics: Heriot-Watt Symposium Volume I
R J Knops

18 Actions of fine abelian groups
C Kosniowski

19 Closed graph theorems and webbed spaces
M De Wilde

20 Singular perturbation techniques applied to integro-differential equations
H Grabmüller

21 Retarded functional differential equations: A global point of view
S E A Mohammed

22 Multiparameter spectral theory in Hilbert space
B D Sleeman

24 Mathematical modelling techniques
R Aris

25 Singular points of smooth mappings
C G Gibson

26 Nonlinear evolution equations solvable by the spectral transform
F Calogero

27 Nonlinear analysis and mechanics: Heriot-Watt Symposium Volume II
R J Knops

28 Constructive functional analysis
D S Bridges

29 Elongational flows: Aspects of the behaviour of model elasticoviscous fluids
C J S Petrie

30 Nonlinear analysis and mechanics: Heriot-Watt Symposium Volume III
R J Knops

31 Fractional calculus and integral transforms of generalized functions
A C McBride

32 Complex manifold techniques in theoretical physics
D E Lerner and P D Sommers

33 Hilbert's third problem: scissors congruence
C-H Sah

34 Graph theory and combinatorics
R J Wilson

35 The Tricomi equation with applications to the theory of plane transonic flow
A R Manwell

36 Abstract differential equations
S D Zaidman

37 Advances in twistor theory
L P Hughston and R S Ward

38 Operator theory and functional analysis
I Erdelyi

39 Nonlinear analysis and mechanics: Heriot-Watt Symposium Volume IV
R J Knops

40 Singular systems of differential equations
S L Campbell

41 N-dimensional crystallography
R L E Schwarzenberger

42 Nonlinear partial differential equations in physical problems
D Graffi

43 Shifts and periodicity for right invertible operators
D Przeworska-Rolewicz

44 Rings with chain conditions
A W Chatters and C R Hajarnavis

45 Moduli, deformations and classifications of compact complex manifolds
D Sundararaman

46 Nonlinear problems of analysis in geometry and mechanics
M Atteia, D Bancel and I Gumowski

47 Algorithmic methods in optimal control
W A Gruver and E Sachs

48 Abstract Cauchy problems and functional differential equations
F Kappel and W Schappacher

49 Sequence spaces
W H Ruckle

50 Recent contributions to nonlinear partial differential equations
H Berestycki and H Brezis

51 Subnormal operators
J B Conway

52 Wave propagation in viscoelastic media
F Mainardi
53 Nonlinear partial differential equations and their applications: Collège de France Seminar. Volume I
H Brezis and J L Lions
54 Geometry of Coxeter groups
H Hiller
55 Cusps of Gauss mappings
T Banchoff, T Gaffney and C McCrory
56 An approach to algebraic K-theory
A J Berrick
57 Convex analysis and optimization
J-P Aubin and R B Vintner
58 Convex analysis with applications in the differentiation of convex functions
J R Giles
59 Weak and variational methods for moving boundary problems
C M Elliott and J R Ockendon
60 Nonlinear partial differential equations and their applications: Collège de France Seminar. Volume II
H Brezis and J L Lions
61 Singular systems of differential equations II
S L Campbell
62 Rates of convergence in the central limit theorem
Peter Hall
63 Solution of differential equations by means of one-parameter groups
J M Hill
64 Hankel operators on Hilbert space
S C Power
65 Schrödinger-type operators with continuous spectra
M S P Eastham and H Kalf
66 Recent applications of generalized inverses
S L Campbell
67 Riesz and Fredholm theory in Banach algebra
B A Barnes, G J Murphy, M R F Smyth and T T West
68 Evolution equations and their applications
F Kappel and W Schappacher
69 Generalized solutions of Hamilton-Jacobi equations
P L Lions
70 Nonlinear partial differential equations and their applications: Collège de France Seminar. Volume III
H Brezis and J L Lions
71 Spectral theory and wave operators for the Schrödinger equation
A M Berthier
72 Approximation of Hilbert space operators I
D A Herrero
73 Vector valued Nevanlinna Theory
H J W Ziegler
74 Instability, nonexistence and weighted energy methods in fluid dynamics and related theories
B Straughan
75 Local bifurcation and symmetry
A Vanderbauwhede
76 Clifford analysis
F Brackx, R Delanghe and F Sommen
77 Nonlinear equivalence, reduction of PDEs to ODEs and fast convergent numerical methods
E E Rosinger
78 Free boundary problems, theory and applications. Volume I
A Fasano and M Primicerio
79 Free boundary problems, theory and applications. Volume II
A Fasano and M Primicerio
80 Symplectic geometry
A Crumeyrolle and J Grifone
81 An algorithmic analysis of a communication model with retransmission of flawed messages
D M Lucantoni
82 Geometric games and their applications
W H Ruckle
83 Additive groups of rings
S Feigelstock
84 Nonlinear partial differential equations and their applications: Collège de France Seminar. Volume IV
H Brezis and J L Lions
85 Multiplicative functionals on topological algebras
T Husain
86 Hamilton-Jacobi equations in Hilbert spaces
V Barbu and G Da Prato
87 Harmonic maps with symmetry, harmonic morphisms and deformations of metrics
P Baird
88 Similarity solutions of nonlinear partial differential equations
L Dresner
89 Contributions to nonlinear partial differential equations
C Bardos, A Damlamian, J I Díaz and J Hernández
90 Banach and Hilbert spaces of vector-valued functions
J Burbea and P Masani
91 Control and observation of neutral systems
D Salamon
92 Banach bundles, Banach modules and automorphisms of C*-algebras
M J Dupré and R M Gillette
93 Nonlinear partial differential equations and their applications: Collège de France Seminar. Volume V
H Brezis and J L Lions
94 Computer algebra in applied mathematics: an introduction to MACSYMA
R H Rand
95 Advances in nonlinear waves. Volume I
L Debnath
96 FC-groups
M J Tomkinson
97 Topics in relaxation and ellipsoidal methods
M Akgül
98 Analogue of the group algebra for topological semigroups
H Dzinotyiweyi
99 Stochastic functional differential equations
S E A Mohammed

100 Optimal control of variational inequalities
V Barbu
101 Partial differential equations and dynamical systems
W E Fitzgibbon III
102 Approximation of Hilbert space operators. Volume II
C Apostol, L A Fialkow, D A Herrero and D Voiculescu
103 Nondiscrete induction and iterative processes
V Ptak and F-A Potra
104 Analytic functions – growth aspects
O P Juneja and G P Kapoor
105 Theory of Tikhonov regularization for Fredholm equations of the first kind
C W Groetsch
106 Nonlinear partial differential equations and free boundaries. Volume I
J I Díaz
107 Tight and taut immersions of manifolds
T E Cecil and P J Ryan
108 A layering method for viscous, incompressible L_p flows occupying R^n
A Douglis and E B Fabes
109 Nonlinear partial differential equations and their applications: Collège de France Seminar. Volume VI
H Brezis and J L Lions
110 Finite generalized quadrangles
S E Payne and J A Thas
111 Advances in nonlinear waves. Volume II
L Debnath
112 Topics in several complex variables
E Ramírez de Arellano and D Sundararaman
113 Differential equations, flow invariance and applications
N H Pavel
114 Geometrical combinatorics
F C Holroyd and R J Wilson
115 Generators of strongly continuous semigroups
J A van Casteren
116 Growth of algebras and Gelfand–Kirillov dimension
G R Krause and T H Lenagan
117 Theory of bases and cones
P K Kamthan and M Gupta
118 Linear groups and permutations
A R Camina and E A Whelan
119 General Wiener–Hopf factorization methods
F-O Speck
120 Free boundary problems: applications and theory, Volume III
A Bossavit, A Damlamian and M Fremond
121 Free boundary problems: applications and theory, Volume IV
A Bossavit, A Damlamian and M Fremond
122 Nonlinear partial differential equations and their applications: Collège de France Seminar. Volume VII
H Brezis and J L Lions
123 Geometric methods in operator algebras
H Araki and E G Effros
124 Infinite dimensional analysis–stochastic processes
S Albeverio
125 Ennio de Giorgi Colloquium
P Krée
126 Almost-periodic functions in abstract spaces
S Zaidman
127 Nonlinear variational problems
A Marino, L Modica, S Spagnolo and M Degiovanni
128 Second-order systems of partial differential equations in the plane
L K Hua, W Lin and C-Q Wu
129 Asymptotics of high-order ordinary differential equations
R B Paris and A D Wood
130 Stochastic differential equations
R Wu
131 Differential geometry
L A Cordero
132 Nonlinear differential equations
J K Hale and P Martinez-Amores
133 Approximation theory and applications
S P Singh
134 Near-rings and their links with groups
J D P Meldrum
135 Estimating eigenvalues with *a posteriori/a priori* inequalities
J R Kuttler and V G Sigillito
136 Regular semigroups as extensions
F J Pastijn and M Petrich
137 Representations of rank one Lie groups
D H Collingwood
138 Fractional calculus
G F Roach and A C McBride
139 Hamilton's principle in continuum mechanics
A Bedford
140 Numerical analysis
D F Griffiths and G A Watson
141 Semigroups, theory and applications. Volume I
H Brezis, M G Crandall and F Kappel
142 Distribution theorems of L-functions
D Joyner
143 Recent developments in structured continua
D De Kee and P Kaloni
144 Functional analysis and two-point differential operators
J Locker
145 Numerical methods for partial differential equations
S I Hariharan and T H Moulden
146 Completely bounded maps and dilations
V I Paulsen
147 Harmonic analysis on the Heisenberg nilpotent Lie group
W Schempp
148 Contributions to modern calculus of variations
L Cesari
149 Nonlinear parabolic equations: qualitative properties of solutions
L Boccardo and A Tesei
150 From local times to global geometry, control and physics
K D Elworthy

151 A stochastic maximum principle for optimal control of diffusions
U G Haussmann
152 Semigroups, theory and applications. Volume II
H Brezis, M G Crandall and F Kappel
153 A general theory of integration in function spaces
P Muldowney
154 Oakland Conference on partial differential equations and applied mathematics
L R Bragg and J W Dettman
155 Contributions to nonlinear partial differential equations. Volume II
J I Díaz and P L Lions
156 Semigroups of linear operators: an introduction
A C McBride
157 Ordinary and partial differential equations
B D Sleeman and R J Jarvis
158 Hyperbolic equations
F Colombini and M K V Murthy
159 Linear topologies on a ring: an overview
J S Golan
160 Dynamical systems and bifurcation theory
M I Camacho, M J Pacifico and F Takens
161 Branched coverings and algebraic functions
M Namba
162 Perturbation bounds for matrix eigenvalues
R Bhatia
163 Defect minimization in operator equations: theory and applications
R Reemtsen
164 Multidimensional Brownian excursions and potential theory
K Burdzy
165 Viscosity solutions and optimal control
R J Elliott
166 Nonlinear partial differential equations and their applications. Collège de France Seminar. Volume VIII
H Brezis and J L Lions
167 Theory and applications of inverse problems
H Haario
168 Energy stability and convection
G P Galdi and B Straughan
169 Additive groups of rings. Volume II
S Feigelstock

Amenable Banach algebras

Jean-Paul Pier

Centre Universitaire de Luxembourg

Amenable Banach algebras

Longman Scientific & Technical

Copublished in the United States with
John Wiley & Sons, Inc., New York

Longman Scientific & Technical
Longman Group UK Limited
Longman House, Burnt Mill, Harlow
Essex CM20 2JE, England
and Associated Companies throughout the world.

Copublished in the United States with
John Wiley & Sons, Inc., 605 Third Avenue, New York, NY 10158

First published 1988

AMS Subject Classifications: 46H25, 46L05, 46L10

ISSN 0269-3674

British Library Cataloguing in Publication Data
Pier, Jean-Paul
Amenable Banach algebras.—(Pitman
research notes in mathematics series;
ISSN 0269–3674; 172).
1. Banach algebras
I. Title
512′.55 QA326

ISBN 0-582-01480-8

Library of Congress Cataloging-in-Publication Data
Pier, Jean-Paul, 1933–
Amenable Banach algebras.
(Pitman research notes in mathematics series,
ISSN 0269-3674 ; 172)
Bibliography: p.
Includes index.
1. Banach algebras. 2. C*-algebras. 3. Von Neumann
algebras. I. Title. II. Series.
QA326.P534 1988 512′.55 87-33910
ISBN 0-470-21066-4 (USA only)

Printed and bound in Great Britain by
Biddles Ltd, Guildford and King's Lynn

Contents

Preface

Among the locally compact groups interest focused on an important class containing all locally compact abelian groups and all compact groups, the class of amenable groups. It may be described by a lot of properties from various mathematical fields; most often these characterizations express some kind of invariance property or quasi-invariance property. For instance, the locally compact group G is amenable if and only if, on the space of bounded continuous functions defined on G, or equivalently, on the space of essentially bounded functions defined on G, with respect to a left Haar measure λ, there exists a mean, i.e., a state, the values of which are invariant by left translations on the group. The class of amenable groups is closed for the passages to closed subgroups, to quotient groups, and also for extensions.

Several properties characterizing amenability of the locally compact group G concern function algebras related to the group, especially the Banach algebra $L^1(G,\lambda)$ called the group algebra of G. An apparently remoted characterization is formulated in terms of cohomology: The locally compact group G is amenable if and only if, for every Banach module X on the Banach algebra $L^1(G,\lambda)$, every continuous derivation of $L^1(G,\lambda)$ into the topological dual space X* of X is inner, i.e., of the form $f \to f\xi-\xi f$ for an element ξ in X*. So one is led to consider the preceding property, in general, on an arbitrary Banach algebra.

As for the case of groups, one has to find equivalent descriptions of this class of algebras, termed again amenable; to a certain extent they reflect the corresponding characterizations of amenable groups. Several problems admit simple solutions on these remarkable Banach algebras. Very precise results are available for amenable C*-algebras and, more specifically, for amenable von Neumann algebras.

We intend to show the fundamental aspects of amenability on Banach algebras, C*-algebras, von Neumann algebras, and to indicate how amenability is placed in the core of important problems.

At least for the questions on general Banach algebras and C*-algebras, we try to sketch the full proofs; we rely only on classical properties for which the reference sources are listed at the top of the bibliography. At the end of each chapter we add comments summarizing results which could not take place in the main text; the historical observations should also provide some complementary information.

We hope the present Notes may lead to further developments of the theory and contribute to a better understanding of the general phenomenon of amenability.

0 Introduction

A. Fundamental notions

In this first section we recall some facts about topological vector spaces, Banach spaces, Hilbert spaces, Banach algebras, and about locally compact groups; they are explained in detail in the different reference sources listed at the top of the bibliography. We also fix some notations which will remain in force in the sequel.

A net is supposed to be directed by $\prec$. Vector spaces are defined over the field of the complexes.

If S is a nonvoid set, one denotes by $\ell^\infty(S)$ the vector space of bounded complex-valued functions on S. If X is a topological vector space, we designate by C(X) the vector space of complex-valued continuous bounded functions on X; $\|\cdot\|$ is the norm of uniform convergence on this space.

Let E and F be two *topological vector spaces*. One denotes by $\mathcal{L}(E,F)$ the vector space of continuous linear mappings of E into F; we put $\mathcal{L}(E) = \mathcal{L}(E,E)$. Every $p \in \mathcal{L}(E)$ for which $p \circ p = p$ is called *projection* in $\mathcal{L}(E)$; $E = p(E) \oplus (id_E - p)(E)$.

We consider the *duality* between two topological vector spaces E and F: On $E \times F$ there exists a nondegenerate bilinear form $\langle .,. \rangle$. If S is a nonvoid subset of E, we denote by $S^\perp$ the *orthogonal* $\{y \in F : (\forall x \in S)\langle x,y\rangle = 0\}$ of S in F. If $\mathfrak{S}$ is a family of subsets in E, the $\mathfrak{S}$-topology of $\mathcal{L}(E,F)$ is given by a fundamental family of neighborhoods of 0, consisting of the subsets of elements f in $\mathcal{L}(E,F)$ for which $f(M) \subset V$, where $M \in \mathfrak{S}$ and V runs over a fundamental family of neighborhoodsof 0_F. In case $\mathfrak{S}$ is the family of finite subsets in E, the topology is called *topology of simple convergence*. The topology $\sigma(E,F)$ [resp. $\sigma(F,E)$] is the coarsest topology on E [resp. F] with respect to which, for any $y \in F$ [resp. any $x \in E$], the mapping $x \mapsto \langle x,y\rangle$ [resp. $y \mapsto \langle x,y\rangle$] is continuous.

Let E be a (complex) Banach space. We denote by E* the topological dual space of E. For $x \in E$ and $f \in E^*$, one expresses the *duality relation*

$$\langle x,f\rangle = f(x);$$

$$\|f\|_{E^*} = \sup \{|\langle x,f\rangle|: x \in E, \|x\|_E \leqq 1\}.$$

We inject E canonically into the topological bidual space E^{**} of E; $\sigma(E,E^*)$ is called the *weak topology* of E and $\sigma(E^*,E)$ is termed the *weak-*-topology* of E^*. The latter is the topology of simple convergence on E^*. If $x \in E$,

$$\|x\|_{E^{**}} = \sup \{|\langle f,x\rangle|: f \in E^*, \|f\|_{E^*} \leqq 1\}$$

$$= \sup \{|\langle x,f\rangle|: f \in E^*, \|f\|_{E^*} \leqq 1\}$$

$$\leqq \sup \{\|f\|_{E^*} \|x\|: \quad f \in E^*, \|f\|_{E^*} \leqq 1\} = \|x\|.$$

With respect to the weak-*-topology, E is dense in E^{**} ([B] Chap. IV, §2, Proposition 5). The Banach space E is called *reflexive* in case $E = E^{**}$.

Let E be a Banach space. If S is a nonvoid subset of E, we denote by $\overline{co}S$ the closure of the convex hull of E, with respect to the topology of E, or, equivalently, with respect to the weak topology of E ([B] Chap. IV, §1, Proposition 2). The closed unit ball of E^* is weak-*-compact ([B] Chap. IV, §2, Proposition 6).

More generally let us consider two Banach spaces E_1 and E_2. On $\mathcal{L}(E_1,E_2)$ one considers the *strong operator topology:* For $T \in \mathcal{L}(E_1,E_2)$ a fundamental family of neighborhoods is given by

$$\{S \in \mathcal{L}(E_1,E_2): \quad (\forall x \in F) \ \|(S-T)(x)\| < \varepsilon\}$$

where F is a finite subset of E, and $\varepsilon \in \underset{\sim}{R}_+^*$. One also defines the *weak operator topology:* For $T \in \mathcal{L}(E_1,E_2)$ a fundamental family of neighborhoods is constituted by

$$\{S \in \mathcal{L}(E_1,E_2): (\forall x \in F_1) \ (\forall y \in F_2) \ |\langle x,(S-T)y\rangle| < \varepsilon\} ,$$

where F_1 [resp. F_2] is a finite subset of E_1 [resp. E_2] and $\varepsilon \in \underset{\sim}{R}_+^*$. The closures of a convex subset in $\mathcal{L}(E_1,E_2)$ with respect to these topologies coincide. If E is a reflexive Banach space, the closed unit ball of E is compact with respect to the weak operator topology ([DS] VI, 1.5, 9.6).

Let E_1 and E_2 be Banach spaces. To $T \in \mathcal{L}(E_1,E_2)$ one associates the transposed operator ${}^tT$: If $x \in E_1$ and $f \in E_2^*$,

$$\langle T(x), f\rangle = \langle x, {}^tT \circ f\rangle;$$

${}^tT \in \mathcal{L}(E_2^*, E_1^*)$.

If E_1 and E_2 are Banach spaces, the operator $T \in \mathcal{L}(E_1,E_2)$ is said to be of *finite rank* in case *Im* T is finite-dimensional. The operator $T \in \mathcal{L}(E_1,E_2)$ is of rank n if and only if there exist linearly independent vectors $\xi_1,\ldots,\xi_n \in E_2$ and $f_1,\ldots,f_n \in E_1^*$ such that

$$T\xi = \sum_{i=1}^{n} f_i(\xi)\xi_i,$$

for every $\xi \in E_1$.

If E is a Banach space consisting of complex-valued functions, that is stable under complex conjugation and contains the unit function 1_E, one calls *mean* on E every $M \in E^*$ for which $M(\bar{f}) = \overline{M(f)}$ whenever $f \in E$, $M(f) \geqq 0$ whenever f is a positive element in E, and $M(1_E) = 1$. We denote by $\mathfrak{M}(E)$ the positive cone of means on E; $\mathfrak{M}(E)$ generates the vector space E^*. Let $\Phi \in E^*$; then $\Phi \in \mathfrak{M}(E)$ if and only if any pair of the following three conditions are satisfied: $\|\Phi\| = 1$; $\Phi \geqq 0$; $\langle 1_E, \Phi\rangle = 1$([P] Proposition 3.2).

Let E_1 and E_2 be Banach spaces. One calls *cross-norm* on the tensor product $E_1 \otimes E_2$ any norm $\|\cdot\|$ on $E_1 \otimes E_2$ such that $\|x_1 \otimes x_2\| = \|x_1\|_{E_1} \|x_2\|_{E_2}$ whenever $x_1 \in E_1$, $x_2 \in E_2$. The completion of $E_1 \otimes E_2$ for a cross-norm $\|\cdot\|_\beta$ is denoted by $E_1 \otimes_\beta E_2$.

One may identify $E_1 \otimes E_2$ with a subspace of $\mathcal{L}(E_1^*,E_2)$ or $\mathcal{L}(E_2,E_1^*)$. If $x = \sum_{i=1}^{n} x_{1,i} \otimes x_{2,i} \in E_1 \otimes E_2$, define $\Phi(x) \in \mathcal{L}(E_1^*,E_2)$ by putting

$$\Phi(x)(f) = \sum_{i=1}^{n} \langle x_{1,i}, f\rangle x_{2,i} \in E_2,$$

for every $f \in E_1^*$. In case $x \neq 0$, one may assume the elements $x_{2,1},\ldots,x_{2,n}$

to be linearly independent. If, for $f \in E_1^*$, there exists $i = 1,\dots,n$ such that $\langle x_{1,i},f\rangle \neq 0$, then $\Phi(x)(f) \neq 0$. Therefore $\Phi(x) \neq 0$. Hence Φ is injective. On the other hand,

$$^t\Phi(x)(g) = \sum_{i=1}^{n} \langle x_{2,i},g\rangle x_{1,i} \in E_1$$

for $x = \sum_{i=1}^{n} x_{1,i} \otimes x_{2,i} \in E_1 \otimes E_2$ and $g \in E_2^*$. If $f \in E_1^*$, $g \in E_2^*$ and $x = \sum_{i=1}^{n} x_{1,i} \otimes x_{2,i} \in E_1 \otimes E_2$, then

$$\langle \Phi(x)(f),g\rangle = \sum_{i=1}^{n} \langle x_{1,i},f\rangle\langle x_{2,i},g\rangle = \langle f, {}^t\Phi(x)(g)\rangle$$

([T] p. 188).

One defines two special cross-norms. If $x = \sum_{i=1}^{n} x_{1,i} \otimes x_{2,i} \in E_1 \otimes E_2$, one puts

$$\begin{aligned}\|x\|_\lambda &= \sup \{|\sum_{i=1}^{n} f(x_{1,i})g(x_{2,i})| : f \in E_1^*, \|f\| \leq 1; g \in E_2^*, \|g\| \leq 1\}\\ &= \sup \{|\langle \Phi(x)(f),g\rangle| : f \in E_1^*, \|f\| \leq 1; g \in E_2^*, \|g\| \leq 1\}\\ &= \|\Phi(x)\| .\end{aligned}$$

The mapping Φ may be extended to an isometry of $E_1 \otimes_\lambda E_2$ into $\mathcal{L}(E_1^*,E_2)$. If $x \in E_1 \otimes E_2$, one puts

$$\|x\|_{\hat{\otimes}} = \|x\|_\gamma = \inf \{\sum_{i=1}^{n} \|x_{1,i}\| \|x_{2,i}\| : x = \sum_{i=1}^{n} x_{1,i} \otimes x_{2,i}\}.$$

The norm $\|\cdot\|_{\hat{\otimes}} = \|\cdot\|_\gamma$ is the greatest cross-norm; the space $E_1 \otimes_\gamma E_2$ is also denoted by $E_1 \hat{\otimes} E_2$. The spaces $(E_1 \otimes_\gamma E_2)^*$, $\mathcal{L}(E_1,E_2^*)$, $\mathcal{L}(E_2,E_1^*)$ are isometric ([T] Chapter IV. Theorem 2.3).

If $f \in (E_1 \otimes_\gamma E_2)^*$, one defines $\Gamma_f \in \mathcal{L}(E_2,E_1^*)$ by putting

$$\langle x_1,\Gamma_f(x_2)\rangle = \langle x_1 \otimes x_2,f\rangle$$

for $x_1 \in E_1$, $x_2 \in E_2$. The mapping $f \mapsto \Gamma_f$ determines a homeomorphism between $(E_1 \otimes_\gamma E_2)^*$, equipped with the weak-*-topology, and $\mathcal{L}(E_2, E_1^*)$, equipped with the topology of simple convergence.

If $\|\cdot\|_\beta$ is a cross-norm on $E_1 \otimes E_2$, one defines the *associated norm* $\|\cdot\|_{\beta^*}$ on $E_1^* \otimes E_2^*$ by putting, for

$$f = \sum_{i=1}^{n} f_{1,i} \otimes f_{2,i} \in E_1^* \otimes E_2^*,$$

$$\|f\|_{\beta^*} = \sup \{|\sum_{i=1}^{n} \sum_{j=1}^{n} \langle x_{1,j}, f_{1,i}\rangle\langle x_{2,j}, f_{2,i}\rangle|:$$

$$x = \sum_{j=1}^{n} x_{1,j} \otimes x_{2,j} \in E_1 \otimes E_2; \ \|x\|_\beta \leqq 1\};$$

$\|\cdot\|_{\beta^*}$ is a cross-norm for $E_1^* \otimes E_2^*$ if and only if $\lambda \leqq \beta \leqq \gamma$ ([T] Chapter IV. Proposition 2.2). Morever, γ^* is the norm λ on $E_1^* \otimes E_2^*$ ([T] Chapter IV. Corollary 2.4).

Let H be a *Hilbert space*. One denotes by $(\cdot|\cdot)$ the corresponding scalar product. If $f \in H^*$, there exists a unique $\eta \in H$ such that $f(\xi) = (\xi|\eta)$ for every $\xi \in H$; thus H^* may be identified with H; a fortiori, H is reflexive. If $f \in \mathcal{L}(H)$, one defines $T^* \in \mathcal{L}(H)$ by putting

$$(T\xi|\eta) = (\xi|T^*\eta)$$

for $\xi, \eta \in H$; in case $T = T^*$, T is said to be *hermitian*. If $A \subset H$, one considers the orthogonal $A^\perp = \{\eta \in H: (\forall\xi \in H)\ (\xi|\eta) = 0\}$ of A. If H_1 and H_2 are closed subspaces of the Hilbert space H, one denotes by $H_1 \ominus H_2$ the orthogonal of H_2 in H_1, i.e., the closed vector space $H_1 \cap H_2^\perp$.

Let T be a hermitian element in the Hilbert space H. If $\eta \in \mathit{Ker}\ T$, then, for every $\xi \in H$, $(T\xi|\eta) = (\xi|T\eta) = 0$; thus $\mathit{Ker}\ T \subset (\mathit{Im}\ T)^\perp$. If $\eta \in (\mathit{Im}\ T)^\perp$, then, for every $\xi \in H$, $0 = (T\xi|\eta) = (\xi|T\eta)$; thus $T\eta = 0$, $\eta \in \mathit{Ker}\ T$ and $(\mathit{Im}\ T)^\perp \subset \mathit{Ker}\ T$. Therefore, $\mathit{Ker}\ T = (\mathit{Im}\ T)^\perp$.

If H_1 and H_2 are Hilbert spaces for which the corresponding scalar products are denoted by $(\cdot|\cdot)_1$, $(\cdot|\cdot)_2$, one defines a scalar product for $H_1 \otimes H_2$ by putting

$$(\xi|\eta) = \sum_{i=1}^{n} \sum_{j=1}^{n} (\xi_{1,i}|\eta_{1,j})_1 (\xi_{2,i}|\eta_{2,j})_2,$$

with $\xi = \sum_{i=1}^{n} \xi_{1,i} \otimes \xi_{2,i}$, $\eta = \sum_{j=1}^{n} \eta_{1,j} \otimes \eta_{2,j} \in H_1 \otimes H_2$. The completion, denoted by $H_1 \otimes H_2$ also, constitutes a Hilbert space ([D) Chapter I, §2, No. 3).

Let H be a Hilbert space. We consider the operations

$$\underset{\mathbb{C} \times H \to H}{(\alpha,\xi) \to \bar{\alpha}\xi}, \quad \underset{H \times H \to H}{(\xi,\eta) \to \xi + \eta}, \quad \underset{H \times H \to H}{(\xi,\eta) \to (\eta|\xi)} .$$

They define the *conjugate Hilbert space* H^c of H; H^c may be identified with the topological dual of H.

Let H be a Hilbert space and let $\{\xi_i: i \in I\}$ be an orthonormal basis of H. One defines a prehilbert structure on the set of all $T \in \mathcal{L}(H)$ such that $\sum_{i\in I} \|T\xi_i\|^2 = \sum_{i,j\in I} |(T\xi_i|\xi_j)|^2 < \infty$ by putting

$$\langle T,T'\rangle_{HS} = \sum_{i\in I} (T\xi_i|T'\xi_i)$$

for two operators T and T' of this type. With respect to the corresponding norm $\|\cdot\|_{HS}$, the completion is the Hilbert space $n(H)$ of the Hilbert-Schmidt operators on H. The norm is independent of the choice of the orthonormal basis. Every $T \in n(H)$ of finite rank is given by $\eta \to \sum_{j=1}^{n} (\eta|\eta_j)\xi_j$, where $\xi_1,\ldots,\xi_n$ are pairwise orthogonal vectors, and $\eta_1,\ldots,\eta_n$ are orthonormal vectors in H; the mapping $T \to \sum_{j=1}^{n} \eta_j \otimes \xi_j$ gives rise to an identification of $n(H)$ with $H^c \otimes H$ ([D] Chapter I, §6, No. 6).

Let A be a *Banach algebra*, i.e., an algebra over the complexes, that is also a Banach space, with

$$\|xy\| \leq \|x\| \|y\|$$

whenever $x,y \in A$. If the algebra admits an identity u_A, it is called *unital;* one supposes that $\|u_A\| = 1$. Every Banach algebra A may be considered to be a closed ideal in a unital Banach algebra $\tilde{A} = A \times \underset{\sim}{C}$. If $(x_1,\alpha_1), (x_2,\alpha_2) \in A \times \underset{\sim}{C}$, $(x_1,\alpha_1) + (x_2,\alpha_2) = (x_1 + x_2,\alpha_1 + \alpha_2)$, $(x_1,\alpha_1)(x_2,\alpha_2) = (x_1x_2 + \alpha_1x_2 + \alpha_2x_1,\alpha_1\alpha_2)$. If $(x,\alpha) \in A \times \underset{\sim}{C}$, $\beta \in \underset{\sim}{C}$, $\beta(x,\alpha) = (\beta x,\beta\alpha)$. The unit of $\tilde{A}$ is $(0_A,1)$. One defines $\|(x,\alpha)\|_{\tilde{A}} = \|x\|_A + |\alpha|$. In the Banach algebra A, (*bounded*) *left* [resp. *right*] *approximate units* are formed by a net $(u_i)_{i\in I}$ such that $\lim \|u_i a-a\| = 0$ [resp. $\lim \|au_i-a\| = 0$] whenever $a \in A$ (and there exists $c \in \underset{\sim}{R}_+$ for which $\|u_i\| \leqq c$ whenever $i \in I$). Left and right approximate units constitute *approximate units*.

Let A again be a Banach algebra. For $a \in A$, one considers the left translation $t_a:x \to ax$ by a in A. One defines

$\bar{\omega} \in \mathcal{L}(A \hat{\otimes} A,A)$ by putting

$$\bar{\omega}(a_1 \otimes a_2) = a_1a_2$$

for $a_1,a_2 \in A$. If $a \in A$ and $f \in A^*$, let $af \in A^*$ be given by

$$\langle x,af\rangle = \langle xa,f\rangle,$$

$x \in A$. For $f \in A^*$ and $F \in A^{**}$, one then defines $F \star f \in A^*$ by

$$\langle a,F \star f\rangle = \langle af,F\rangle,$$

$a \in A$. If $F_1,F_2 \in A^*$, let also

$$\langle f,F_1 \odot F_2\rangle = \langle F_2 \star f,F_1\rangle,$$

$f \in A^*$; then $F_1 \odot F_2 \in A^{**}$. With respect to the ordinary addition, and the multiplication, called *Arens multiplication*, defined by $\odot$, A^{**} constitutes a Banach algebra.

An algebra A is called *involutive* in case it admits a mapping (involution)

$$\begin{aligned} x &\mapsto x^* \\ A &\to A \end{aligned}$$

satisfying the following conditions:

(a) $(x^*)^* = x$ whenever $x \in A$;

(b) $(x+y)^* = x^* + y^*$ and $(xy)^* = y^*x^*$ whenever $x,y \in A$;

(c) $(\alpha x)^* = \bar{\alpha} x^*$ whenever $\alpha \in \underset{\sim}{C}$ and $x \in A$.

Every element x in A such that $x = x^*$ is said to be *hermitian*. We denote by A_h the set of all hermitian elements in A. An element p in A_h such that $p^2 = p$ is called *projection*. If A is unital, we choose the unit to be hermitian. In the latter case, any element a in A such that $a^{-1} = a^*$ is called *unitary*. The set of all unitary elements constitutes the unitary group of A; we denote it by A_u.

In an involutive Banach algebra A we suppose that $\|x^*\| = \|x\|$ for every $x \in A$.

Any closed subalgebra of an involutive Banach algebra A, that is invariant under the involution, is called **-subalgebra* of A.

Let A be an involutive Banach algebra. The linear functional f on A is said to be *hermitian* if $f = f^*$, where $f^*(x) = \overline{f(x^*)}$ for $x \in A$. Every linear functional f on A admits a unique decomposition $f = f_1 + if_2$, where f_1, f_2 are hermitian linear functionals on A; $f_1 = \frac{1}{2}(f+f^*)$, $f_2 = \frac{1}{2i}(f-f^*)$.

The involutive Banach algebra A is called *C*-algebra* if

$$\|x^*x\| = \|x\|^2$$

whenever $x \in A$. The algebra $\tilde{A}$ associated to a C*-algebra A is itself a C*-algebra, the norm being defined by

$$\|x\|_{\tilde{A}} = \sup \{\|xy\|_A : y \in A, \|y\|_A \leq 1\},$$

$x \in \tilde{A}$([T] Chapter I. Proposition 1.5).

Let H be a Hilbert space; $\mathcal{L}(H)$ is a C*-algebra in which the involutive image T^* of $T \in \mathcal{L}(H)$ is defined by

$$(T\xi|\eta) = (\xi|T^*\eta)$$

for $\xi, \eta \in H$. The C*-algebra $\mathcal{L}(H)$ admits the unit id_H. If $T \in \mathcal{L}(H)$ is hermitian, then

$$\|T\| = \sup \{|(T\xi|\xi)|: \ \xi \in H, \ \|\xi\| \leq 1\}.$$

Any C*-algebra admits approximate units bounded by 1. If, in particular, A is unital, then

$$\|x\| = \|x^*x\|^{\frac{1}{2}} = 1$$

whenever $x \in A_u$. To every C*-algebra A corresponds a Hilbert space H such that A is isomorphic to a C*-subalgebra of $\mathcal{L}(H)$.

If A is a C*-algebra, the Banach algebra A^c deduced from A by adopting the opposite multiplication, is also a C*-algebra for the same Banach space and the same involution; the image of $a \in A$ in A^c is denoted by a^c.

Let A be a C*-algebra. If $x \in A$, x^*x is called *positive element* in A. We denote by A_+ the positive cone of positive elements in A([T] Chapter I. Theorem 6.1); $A_+ \subset A_h$. For $x_1,x_2 \in A_h$, we write $x_1 \leq x_2$ in case $x_2-x_1 \in A_+$. If $a,b,x \in A$ and $a \leq b$, then also $x^*ax \leq x^*bx$ ([D'] 1.6.8). For two projections p_1 and p_2 in A, one has $p_1 \leq p_2$ if and only if $p_1p_2 = p_2p_1 = p_1$. The linear functional f on A is called *positive* if $f(x^*x) \geq 0$ whenever $x \in A$. The positive linear functional is necessarily continuous. If f is any continuous linear functional on A that is hermitian, there exists a unique pair of positive (continuous) linear functionals f_+, f_- on A such that $f = f_+ - f_-$ and $\|f\| = \|f_+\| + \|f_-\|$ ([D'] 2.6.4; 12.3.4). The positive linear functional f on the C*-algebra A is called *normal* if, for every increasing net (x_i) in A_+ admitting the least upper bound x, $(f(x_i))$ admits the least upper bound f(x).

Let H be a Hilbert space. If S is a nonvoid subset of $\mathcal{L}(H)$, we consider the set I(S) of all *isometries* in S, i.e., all $U \in S$ such that $\|U\xi\| = \|\xi\|$ whenever $\xi \in H$; $\|U^*\| = \|U\| = 1$. Let A be an involutive subalgebra of $\mathcal{L}(H)$ and let $a \in I(A)$. For all $\xi \in H$

$$\|(a^*a-\mathrm{id}_H)\xi\|^2 = (a^*a\xi|a^*a\xi) - (a^*a\xi|\xi) - (a^*a\xi|\xi) + (\xi|\xi)$$

$$\leq \|a^*\|^2 \ \|a\xi\|^2 - \|a\xi\|^2 - \|a\xi\|^2 + \|\xi\|^2 = 0.$$

Hence $a^*a = \mathrm{id}_H$. Moreover, $(aa^*)(aa^*) = a(a^*a)a^* = aa^*$. The operator T in $\mathcal{L}(H)$ is called *partial isometry* if there exists a (closed) vector space $H^{(T)}$ in H such that $\|T\xi\| = \|\xi\|$ whenever $\xi \in H^{(T)}$ and $T\xi = 0$ whenever $\xi \in H^{(T)\perp}$. As a partial isometry the element T in $\mathcal{L}(H)$ is characterized by each of the following conditions:

(a) T^*T is a projection; (b) TT^* is a projection; (c) $T^*TT^* = T^*$. Then one also has $(T^*T\xi|\eta) = (\xi|\eta)$ for all $\xi, \eta \in H^{(T)}$; T^* is a partial isometry as well, and $T^*(H) \subset H^{(T)}$.

Let H be a Hilbert space. If S is a nonvoid subset of $\mathcal{L}(H)$, we denote by $S^{\sim}$ the *commutant* of S, i.e., the set of all $T \in \mathcal{L}(H)$ such that $TU = UT$ whenever $U \in S$. We have $S^{\sim\sim} \subset S$ and $S^{\sim\sim\sim} = S^{\sim}$; $S^{\sim\sim}$ is the *bicommutant* of S.

In a group G the identity element is denoted by e.

A *topological group* is a group G equipped with a topology for which the mappings $\begin{matrix}(x,y) \mapsto xy \\ G \times G \to G\end{matrix}$ and $\begin{matrix}x \mapsto x^{-1} \\ G \to G\end{matrix}$ are continuous. If $f: G \to \mathbb{C}$ and $a \in G$, let ${}_af(x) = f(ax)$, $f_a(x) = f(xa)$, $x \in G$. One considers the Banach space $RUC(G)$ of the complex-valued functions f, that are bounded on G, and right uniformly continuous, i.e., for every $\varepsilon > 0$, there exists a neighborhood V of e such that $|f(x) - f(y)| < \varepsilon$ whenever $x,y \in G$ with $xy^{-1} \in V$.

If the group G is *locally compact*, one considers the vector space $C_0(G)$ of the complex-valued continuous (bounded) functions on G that vanish at infinity, i.e., if $\varepsilon > 0$, there exists a compact subset K in G such that $|f(x)| < \varepsilon$ whenever $x \in G \setminus K$. The locally compact group G admits a left *Haar measure* $\lambda = dx$ that is positive and left invariant, i.e., for all $f \in L^1(G,\lambda) = L^1(G)$ and $a \in G$,

$$\int_G {}_af(x)dx = \int_G f(x)dx.$$

The locally compact group G admits a modular function Δ, i.e., a continuous homomorphism of G into the multiplicative group $\mathbb{R}_+^*$ such that, for all $f \in L^1(G)$ and $a \in G$,

$$\int_G f(xa)\,\Delta(a)dx = \int_G f(x)dx.$$

If $\Delta = 1$, the group is said to be *unimodular*. Discrete groups, locally compact abelian groups, compact groups are unimodular. We denote by $M^1(G)$ the Banach algebra of bounded measures on G, the multiplication being the *convolution product* $*$ defined by

$$\int_G f(z)d(\mu*\nu)(z) = \int_G \int_G f(xy)d\mu(x)d\nu(y)$$

for $\mu, \nu \in M^1(G)$ and $f \in C_o(G)$. If $a \in G$, one denotes by δ_a the Dirac measure in $M^1(G)$ concentrated at a. In $M^1(G)$, $L^1(G)$ is a Banach subalgebra called group algebra; the corresponding norm is denoted by $\|\cdot\|_1$. For $f,g \in L^1(G)$,

$$f * g(x) = \int_G f(y)g(y^{-1}x)dy,$$

$x \in G$. In case G is discrete, we write $\ell^1(G)$ for the group algebra; it consists of the functions $u = \sum_{n\in\underset{\sim}{N}^*} \alpha_n\delta_{x_n}$, where (α_n) is a sequence in $\underset{\sim}{C}$, (x_n) is a sequence in G, and $\|u\|_1 = \sum_{n\in\underset{\sim}{N}^*} |\alpha_n| < \infty$.

B. Cohomology and amenable groups

Let A be a Banach algebra and let G be a topological group. We say that the Banach space X constitutes a *left Banach* A-*module* [resp. G-*module*] if there exists a mapping

$$(a,\xi) \mapsto a\xi$$

$$A \times X \to X$$

$$[\text{resp.}(x,\xi) \mapsto x\xi$$

$$G \times X \to X]$$

satisfying the following properties:

(i) X is a left A-module

$(\forall a \in A)\ (\forall b \in A)\ (\forall \xi \in X)\quad (a + b)\xi = a\xi + b\xi,$

$(\forall a \in A)\ (\forall \xi \in X)\ (\forall \eta \in X)\ a(\xi + \eta) = a\xi + a\eta,$

$$(\forall \alpha \in \underset{\sim}{C})\ (\forall a \in A)\ (\forall \xi \in X) \quad (\alpha a)\xi = \alpha(a\xi) = a(\alpha\xi),$$

$$(\forall a \in A)\ (\forall b \in A)\ (\forall \xi \in X) \quad (ab)\xi = a(b\xi)$$

[resp. X is a left G-module

$$(\forall x \in G)\ (\forall \xi \in X)\ (\forall \eta \in X) \quad x(\xi + \eta) = x\xi + x\eta,$$

$$(\forall \alpha \in \underset{\sim}{C})\ (\forall x \in G)\ (\forall \xi \in X) \quad \alpha(x\xi) = x(\alpha\xi),$$

$$(\forall x \in G)\ (\forall y \in G)\ (\forall \xi \in X) \quad (xy)\xi = x(y\xi)].$$

(ii) The mapping $\underset{A \times X \to X}{(a,\xi) \mapsto a\xi}$ is continuous, i.e.,

$$(\exists k \in \underset{\sim}{R}_+^*)\ (\forall a \in A)\ (\forall \xi \in X) \quad \|a\xi\|_X \leqq k\,\|a\|_A\ \|\xi\|_X$$

$$[\text{resp. } (\exists k \in \underset{\sim}{R}_+^*)\ (\forall x \in G)\ (\forall \xi \in X) \quad \|x\xi\|_X \leqq k\ \|\xi\|_X$$

and, for every $\xi \in X$, the mapping $\underset{G \to X}{x \mapsto x\xi}$ is continuous].

In the same manner, one defines a *right Banach* A-*module* [resp. G-*module*]. The (bilateral) *Banach* A-*module* [resp. G-*module*] is a left and right A-module [resp. G-module] for which

$$(\forall a \in A)\ (\forall b \in A)\ (\forall \xi \in X) \quad a(\xi b) = (a\xi)b$$

$$[\text{resp. } (\forall x \in G)\ (\forall y \in G)\ (\forall \xi \in X) \quad x(\xi y) = (x\xi)y].$$

Every Banach algebra A may be considered to be a Banach A-module (with $k = 1$). If A is a Banach algebra and X is a left [resp. right] Banach A-module, one may define canonically a right [resp. left] Banach A-module structure on X* by putting

$$\langle \xi, Fa \rangle = \langle a\xi, F \rangle \ [\text{resp. } (\xi, aF\rangle = \langle \xi a, F \rangle]$$

for $a \in A$, $\xi \in X$, $F \in X^*$; then one has

$$\|Fa\| = \sup\{|\langle a\xi, F\rangle| : \xi \in X,\ \|\xi\|_X \leqq 1\} \leqq \|F\|\ k\ \|a\|_A$$

$$[\text{resp. } \|aF\| = \sup\{|\langle \xi a, F\rangle| : \xi \in X,\ \|\xi\|_X \leqq 1\} \leqq \|F\|\ k\ \|a\|_A].$$

If G is a topological group and X is a left [resp. right] Banach G-module, one defines similarly a right [resp. left] G-module structure on X^* by putting

$$\langle\xi, Fx\rangle = \langle x\xi, F\rangle \text{ [resp. } \langle\xi, xF\rangle = \langle\xi x, F\rangle],$$

$x \in G$, $\xi \in X$, $F \in X^*$; if $F \in X^*$, $x \mapsto Fx$ [resp. $x \mapsto xF$] maps G into X^* and is continuous for the weak-*-topology on X^*.

In the sequel we always suppose norms on modules to be normalized in order to have $k = 1$ in the preceding set-up.

Let X and Y be Banach modules over the same Banach algebra A. The element $f \in \mathcal{L}(x,y)$ is called *Banach* A-*module homomorphism* if

$$af(\xi) = f(a\xi)$$

$$f(\xi)a = f(\xi a)$$

whenever $a \in A$ and $\xi \in X$.

Let A be a Banach algebra. If X is a Banach A-module, every closed subspace Y of X that is a Banach A-module for the induced structure, is called *Banach* A-*submodule* of X. If $a \in A$ and $\xi, \xi' \in X$ such that $\xi - \xi' \in Y$, then $a\xi - a\xi' = a(\xi - \xi') \in Y$. Thus one may define a Banach A-module structure on the Banach quotient space by putting $a\dot{\xi} = \dot{\widehat{a\xi}}$ for $a \in A$, $\xi \in X$, $\dot{\xi} \in X/Y$.

Let A and B be Banach algebras. If $a \in A$, $b \in B$ and $x \in A$, $y \in B$, we consider

$$\sigma_a(x \otimes y) = a(x \otimes y) = ax \otimes y, \quad \tau_b(x \otimes y) = (x \otimes y)b = x \otimes yb$$

and

$$\sigma'_b(x \otimes y) = b.(x \otimes y) = x \otimes by, \quad \tau'_a(x \otimes y) = (x \otimes y).a = xa \otimes y.$$

If A is a Banach algebra, the mappings σ_a, τ_a ($a \in A$) and also the mappings σ'_a, τ'_a ($a \in A$) define on $A \hat{\otimes} A$ a Banach A-module structure; these structures commute. In principle, one studies the first of these structures.

If A is a Banach algebra and X is a Banach A-module, we denote by $Z(A,X)$ the closed vector subspace $\{\xi \in X : (\forall a \in A)\ a\xi = \xi a\}$ of X; this subspace is invariant by the action of any continuous endomorphism commuting with the action of A.

Let A be a Banach algebra and let X be a Banach A-module. We define a continuous linear mapping $d_1 : X \to \mathcal{L}(A,X)$ putting

$$d_1\xi(a) = a\xi - a\xi$$

for $a \in A$ and $\xi \in X$. We define a continuous linear mapping d_2 of $\mathcal{L}(A,X)$ into the space $\mathcal{L}_2(A,X)$ of the continuous bilinear mappings of $A \times A$ into X; we put

$$d_2 F(a,b) = aF(b) - F(ab) + F(a)b$$

for $F \in \mathcal{L}(A,X)$ and $(a,b) \in A \times A$. For all $(a,b) \in A \times A$ and $\xi \in X$, we have $d_2 d_1 \xi(a,b) = 0$; hence $Im\ d_1 \subset Ker\ d_2$. The vector space $H_1(A,X) = Ker\ d_2/Im\ d_1$ constitutes the *first cohomology group* of A on X.

The linear mapping $D: A \to X$ satisfying

$$D(ab) = aD(b) + D(a)b$$

whenever $a,b \in A$, is called *derivation; Ker* d_2 is the set of all continuous derivations of A into X. For every $\xi \in X$, the mapping

$$d_1\xi : a \mapsto a\xi - a\xi$$

$$A \to A$$

is termed *inner derivation* of A into X. The condition $H_1(A,X) = \{0\}$ expresses that every continuous derivation of A into X is inner.

Any derivation of a C*-algebra is automatically continuous ([S] 4.1.3).

If G is a topological group and X is a Banach G-module, one considers similarly the space Z(G,X*) of all *continuous derivations* of G into X*, i.e., all mappings D from G into X* that are norm-bounded, continuous with respect to the weak-*-topology of X* and satisfy the condition

$$D(xy) = xD(y) + D(x)y$$

whenever $x,y \in G$. We denote by $N(G,X^*)$ the subspace of all *continuous inner derivations*, i.e., all mappings $d_1\xi (\xi \in X^*)$ defined by

$$d_1\xi(x) = x\xi - \xi x ,$$

$x \in G$. Let $H_1(G,X^*) = Z(G,X^*)/N(G,X^*)$.

We consider a Banach algebra A with unit e [resp. a topological group G with identity element e] and a Banach A-module [resp. G-module] X. We say that A [resp. G] acts as a *left* (or *right*) *zero* on X [resp. X^*] if $a\xi = 0$ (or $\xi a = 0$) whenever $a \in A$ [resp. $a \in G$] and $\xi \in X$ [resp. $\xi \in X^*$]. Then we have $H_1(A,X) = \{0\}$ [resp. $H_1(G,X^*) = \{0\}$]. As a matter of fact, let us consider the left action. If D is any continuous derivation of A into X [resp. G into X^*], put $\eta_0 = -D(e)$. For every $a \in A$ [resp. $a \in G$], we have

$$D(a) = D(ea) = D(e)a = a\eta_0 - \eta_0 a,$$

i.e., the derivation is inner.

PROPOSITION 0.1: If A is a Banach algebra with unit e [resp. G is a topological group with identity element e] and X is a Banach A-module [resp. G-module], then $H_1(A,X)$ [resp. $H_1(G,X^*)$] is isomorphic to $H_1(A, exe)$ [resp. $H_1(G,eX^*e)$].

PROOF: Let $Y = X$ [resp. X^*] and consider $\ell: \underset{Y \to Y}{\xi \mapsto e\xi}$, $r: \underset{Y \to Y}{\xi \mapsto \xi e}$. One readily verifies that Y is the direct sum of the Banach A-submodules [resp. G-submodules] $eYe = \ell \circ r(Y)$, $Y_1 = (id_Y - r) \circ \ell(Y)$, $Y_2 = (id_Y - \ell) \circ r(Y)$, $Y_3 = (id_Y - \ell) \circ (id_Y - r)(Y)$ that are images of Y by pairwise commuting projections; $H_1(A,X)$ [resp. $H_1(G,X^*)$] is the direct sum of the cohomology groups associated to eYe, Y_1, Y_2, Y_3. The algebra A [resp. the group G] acts as a left zero on Y_2, Y_3, a right zero on Y_1. Hence $H_1(A,X) \simeq H_1(A,eXe)$ [resp. $H_1(G,X^*) \simeq H_1(G,eX^*e)$]. □

Let A be a Banach algebra with unit e [resp. G a topological group with identity element e] and let X be a Banach A-module [resp. G-module]; X is called *unital* if $e\xi e = \xi$ whenever $\xi \in X$. That situation holds exactly in case $e\xi = \xi = \xi e$ whenever $\xi \in X$. For every Banach A-module [resp. G-module] X, eXe is unital. If D is a derivation of A [resp. G] into a unital Banach module, then $D(e) = D(e^2) = eD(e) + D(e)e = D(e) + D(e)$, hence $D(e) = 0$.

Consider a unital Banach algebra A [resp. a topological group G]. By Proposition 0.1, if $H_1(A,Y) = \{0\}$ for every unital Banach A-module Y [resp. $H_1(G,Y^*) = \{0\}$ for every unital Banach G-module Y], then $H_1(A,X) = \{0\}$ for every Banach A-module X [resp. $H_1(G,X^*) = \{0\}$ for every Banach G-module X].

If A is a Banach algebra and X is a Banach A-module, one may equip X with a Banach $\tilde{A}$-module structure by putting $u\xi = \xi = \xi u$ for $\xi \in X$, u being the unit in $\tilde{A}$. On X* the induced structure coincides with the Banach $\tilde{A}$-module structure associated to the given Banach A-module structure.

PROPOSITION 0.2: If A is a Banach algebra and X is a Banach A-module, then $H_1(A,X) \simeq H_1(\tilde{A},X)$.

PROOF: Let u be the unit of $\tilde{A}$ and consider a continuous derivation of $\tilde{A}$ into the Banach $\tilde{A}$-module X. Then $D(u) = 0$. The canonical injection of A into $\tilde{A}$ induces an isomorphism of $H_1(\tilde{A},X)$ onto $H_1(A,X)$. □

Consider a Banach algebra A and a Banach A-module X. We say that X is *essential* if X coincides with $AX = \{a\xi : a \in A, \xi \in X\}$ and $XA = \{\xi a : a \in A, \xi \in X\}$.

Every unital Banach module is essential.

Let A be a Banach algebra admitting the approximate units $(u_i)_{i\in I}$ and let X be an essential Banach A-module. If $\xi \in X$, there exist $a_1, a_2 \in A$ and $\xi_1, \xi_2 \in X$ such that $a_1\xi_1 = \xi = \xi_2 a_2$. For every $i \in I$,

$$\|u_i\xi-\xi\| \leq \|u_i a_1 - a_1\| \; \|\xi_1\|, \|\xi u_i - \xi\| \leq \|a_2 u_i - a_2\| \; \|\xi_2\| .$$

Hence $\lim \|u_i\xi-\xi\| = 0 = \lim \|\xi u_i - \xi\|$.

PROPOSITION 0.3: Let A be a Banach algebra admitting bounded approximate units and let X be a Banach A-module. Consider $X_1 = \{a\xi b : a \in A, b \in A, \xi \in X\}$. Then $H_1(A,X^*) \simeq H_1(A,X_1^*)$.

PROOF: Let also $X_2 = \{a\xi : a \in A,\ \xi \in X\}$. As A admits bounded approximate units (u_i), Cohen's classical factorization theorem ([40] 32.22) shows that X_2 and X_1 are Banach A-submodules in X.

The Banach algebra A acts as a left zero on X/X_2; therefore

$$H_1(A, (X/X_2)^*) = \{0\}, \text{ i.e.,}$$

$$H_1(A, X_2^{\perp}) = \{0\}. \tag{1}$$

As $\mathcal{L}(X^*) \simeq (X \otimes_{\gamma} X^*)^*$, there exist a subnet (v_s) of (u_i) and $F \in \mathcal{L}(X^*)$ such that, in the topology $\sigma((X \otimes_{\gamma} X^*)^*, X \otimes_{\gamma} X^*)$,

$$\langle \xi, F(f)\rangle = \lim_s \langle \xi, f\, v_s\rangle$$

whenever $\xi \in X$, $f \in X^*$. If $a \in A, \xi \in X$, $f \in X^*$, we have

$$\langle a\xi, F(f)\rangle = \lim_s \langle a\xi, f v_s\rangle = \lim_s \langle v_s a\xi, f\rangle = \langle a\xi, f\rangle.$$

Thus $(id_{X^*} - F)(X^*) = X_2^{\perp}$. If $\xi \in X$, $f \in X_2^{\perp}$, we have

$$\langle \xi, F(f)\rangle = \lim_s \langle v_s\ \xi, f\rangle = 0.$$

Therefore $(id_{X^*} - F)|_{X_2^{\perp}} = id_{X_2^{\perp}}$ and $id_{X^*} - F$ is the projection of X^* onto $X_2^{\perp}$. As $X^*/X_2^{\perp} \simeq X_2^*$, we deduce now from (1) that

$$H_1(A, X_2^*) \simeq H_1(A, X^*).$$

Similarly, one establishes then that

$$H_1(A, X_1^*) \simeq H_1(A, X_2^*). \qquad \square$$

If G is a locally compact group, there exists a one-to-one correspondence between the unital Banach G-modules and the essential Banach L'(G)-modules ([P] p. 101-102).

The *locally compact group* G is defined to be amenable if, on anyone of the spaces $E = L^{\infty}(G,\lambda)$, $C(G)$, $\mathcal{RUC}(G)$, it admits a *mean* M that is *left invariant*, i.e., $M({}_af) = M(f)$ whenever $f \in E$ and $a \in G$. Then E admits a *mean* M that is *biinvariant* (i.e., left and right invariant) : $M({}_af_b) = M(f)$ whenever $f \in E$ and $a,b \in G$. The simplest examples of amenable groups are provided by the compact groups and the locally compact abelian groups, or more generally, the locally compact solvable groups. The discrete free group on two generators is nonamenable.

Amenability of the locally compact group G may be characterized by anyone of the following properties ([P] Theorem 11.8):

(i) For every unital Banach G-module X, $H_1(G,X^*) = \{0\}$.

(ii) For every Banach G-module X, $H_1(G,X^*) = \{0\}$.

(iii) For every essential Banach $L^1(G)$-module X, $H_1(L^1(G),X^*) = \{0\}$.

(iv) For every Banach $L^1(G)$-module X, $H_1(L^1(G),X^*) = \{0\}$.

(v) $H_1(G,(\mathcal{RUC}(G)/\mathbb{C}\underset{\sim}{1})^*) = \{0\}$ with $xf = f$, $fx = {}_xf$ for $f \in \mathcal{RUC}(G)$, $x \in G$.

Notes

The pioneering work concerning the characterization of the amenable locally compact groups in the language of cohomology has been realized by Johnson [42]. He proves 0.1 to 0.3 as well as the equivalence of conditions (i) - (v) with the amenability of the group; he gives credit to Ringrose for the proof of the sufficiency of (v) for amenability. In this context, see [46].

Another proof of the characterization of amenability for a locally compact group by property (iv) is due to Khelemskii and Sheinberg [50]. See also Guichardet [33].

1 Amenability of the Banach algebras

We begin by formulating the fundamental definition.

DEFINITION 1.1: The Banach algebra A is called amenable if for every Banach A-module X, one has $H_1(A,X^*) = \{0\}$.

Thus, the locally compact group G is amenable if and only if the Banach algebra $L^1(G)$ is amenable.

We first establish a collection of properties equivalent to Definition 1.1. We then examine the class of amenable Banach algebras. We list general examples of amenable Banach algebras and give indications on problems related to the amenability of Banach algebras.

A. Properties and characterizations of amenable Banach algebras

For every locally compact group G, the group algebra $L^1(G)$ admits bounded approximate units. Moreover, to every locally compact group G corresponds a particular Banach algebra, the Fourier algebra A(G). Amenability of G is characterized by the existence of bounded approximate units in A(G) ([P] 10.B. Theorem 10.4).

LEMMA 1.2: Let A be a Banach algebra. On A** one considers the Banach A-module structure defined by

$$a\Phi = {}^{tt}t_a\Phi, \quad \Phi a = 0$$

for $a \in A$, $\Phi \in A^{**}$. If $H_1(A,A^{**}) = \{0\}$, then A admits bounded right approximate units.

PROOF: The canonical injection θ of A into A** is a continuous derivation. By hypothesis, there exists $\Phi \in A^{**}$ such that $\theta = d_1\Phi$. One may determine a net $(u_i)_{i \in I}$, bounded by $c > 0$, in A, such that $\lim \theta(u_i) = \Phi$ in the

$\sigma(A^{**},A^*)$-topology. For all $a \in A$ and $f \in X^*$,

$$\langle a,f\rangle = \langle f,\theta(a)\rangle = \langle f,a\Phi-\Phi a\rangle = \langle f,a\Phi\rangle$$

$$= \langle fa,\Phi\rangle = \lim\langle u_i,fa\rangle = \lim\langle au_i,f\rangle.$$

Let $F = \{a_1,\ldots,a_n\}$ be a finite subset of A and $\varepsilon \in \underset{\sim}{R}_+^*$. There exists a convex linear combination $v_{F,\varepsilon}$ of $\{u_i : i \in I\}$, bounded by c, such that $\|a_j v_{F,\varepsilon} - a_j\| < \varepsilon$ whenever $j = 1,\ldots,n$. Denote by $\mathcal{F}$ the set of all finite subsets of A. The family $(v_{F,\varepsilon})_{F\in\mathcal{F},\varepsilon\in\underset{\sim}{R}_+^*}$ constitutes bounded approximate units where, for (F,ε), $(F',\varepsilon') \in \mathcal{F} \times \underset{\sim}{R}_+^*$, one has $v_{F,\varepsilon} < v_{F',\varepsilon'}$ if and only if $F \subset F'$ and $\varepsilon' < \varepsilon$. □

PROPOSITION 1.3: Every amenable Banach algebra admits bounded right, left, bilateral approximate units.

PROOF: The statement follows readily from Lemma 1.2 and the corresponding left-hand version. Notice that if $(u_i)_{i\in I}$ forms left approximate units, bounded by c, and $(u'_{i'})_{i'\in I'}$ forms right approximate units, bounded by c', then $(u_i + u'_{i'} - u_i u'_{i'})_{(i,i')\in I\times I'}$ constitutes approximate units, bounded by $c + c' + cc'$. □

DEFINITION 1.4: Let A be a Banach algebra. The bounded net(u_i) in $A \hat{\otimes} A$ is called approximate diagonal of A if $\lim(u_i a - au_i) = 0$ and $\lim \bar{\omega}(u_i)a = a$ whenever $a \in A$.

With respect to the Banach A-module structure of $A \hat{\otimes} A$ we have then $\lim (\bar{\omega}(u_i)a - a\bar{\omega}(u_i)) = 0$ whenever $a \in A$, so also $\lim a\bar{\omega}(u_i) = a$ whenever $a \in A$. Therefore the family $(\bar{\omega}(u_i))$ constitutes bounded approximate units in A.

DEFINITION 1.5: Let A be a Banach algebra. The element Φ of $(A \hat{\otimes} A)^{**}$ is called virtual diagonal of A if $\Phi a = a\Phi$ and ${}^{tt}\bar{\omega}(\Phi)a = a$ whenever $a \in A$.

LEMMA 1.6: The Banach algebra A admits an approximate diagonal if and only if it admits a virtual diagonal.

PROOF: (1) Every approximate diagonal admits, in $(A \hat{\otimes} A)^{**}$, a weak-*-limit point which constitutes a virtual diagonal.

(2) Assume the existence of a virtual diagonal Φ. One may determine a bounded net $(u_i)_{i\in I}$ in $A \hat{\otimes} A$ that converges to Φ with respect to the $\sigma((A \hat{\otimes} A)^{**}, (A \hat{\otimes} A)^*)$-topology. Then, for every $a \in A$, $(au_i - u_i a)$ converges to 0 in the weak topology of $A \hat{\otimes} A$, and $(\bar{\omega}(u_i)a)$ converges to a in the weak topology of A.

Let $F = \{a_1,\ldots,a_n\} \subset A$ and $\varepsilon \in \underset{\sim}{R}^*_+$. The bounded net $(a_1u_i - u_ia_1, \bar{\omega}(u_i)a_1, \ldots, a_nu_i - u_ia_n, \bar{\omega}(u_i)a_n)_{i\in I}$ converges to $(0,a_1,\ldots,0,a_n)$ in $((A \hat{\otimes} A) \oplus A)^n$ equipped with the weak product topology. There exists then a convex linear combination $v_{F,\varepsilon}$ of elements from $\{u_i : i \in I\}$ such that $\|a_jv_{F,\varepsilon} - v_{F,\varepsilon}a_j\| < \varepsilon$ and $\|\bar{\omega}(v_{F,\varepsilon})a_j - a_j\| < \varepsilon$ whenever $j = 1,\ldots,n$. Denote by $\mathcal{F}$ the set of finite subsets in A. The net $(v_{F,\varepsilon})_{F\in\mathcal{F},\ \varepsilon\in\underset{\sim}{R}^*_+}$ constitutes an approximate diagonal where again, for $(F,\varepsilon), (F',\varepsilon') \in \mathcal{F} \times \underset{\sim}{R}^*_+, v_{F,\varepsilon} \prec v_{F',\varepsilon'}$ if and only if $F \subset F'$, $\varepsilon' < \varepsilon$. □

PROPOSITION 1.7: The Banach algebra A is amenable if and only if it admits an approximate diagonal or, equivalently, a virtual diagonal.

PROOF: (1) We assume the existence of a virtual diagonal Φ which is a weak-*-limit point for a net (u_i) in $A \hat{\otimes} A$ such that the net $\bar{\omega}(u_i)$ constitutes bounded approximate units in A. In order to establish the amenability of A, by Proposition 0.3 it suffices to show that, if X is any essential Banach A-module X, any continuous derivation D of A into X* is inner.

If $\xi \in X$, we define $F_\xi \in (A \hat{\otimes} A)^*$ by putting

$$F_\xi(a \otimes b) = \langle \xi, aD(b) \rangle$$

for $a,b \in A$; $\|F_\xi\| \leq \|D\| \; \|\xi\|$. We consider

$$f : \xi \mapsto \langle F_\xi, \Phi \rangle;$$

$f \in X^*$. For all $a, b, c \in A$ and $\xi \in X$,

$$\langle b \otimes c, F_\xi a - aF_\xi\rangle + \langle \xi, bc\, D(a)\rangle$$

$$= \langle a(b \otimes c), F_\xi\rangle - \langle (b \otimes c)a, F_\xi\rangle + \langle \xi, bc\, D(a)\rangle$$

$$= \langle ab \otimes c, F_\xi\rangle - \langle b \otimes ca, F_\xi\rangle + \langle \xi, bc\, D(a)\rangle$$

$$= \langle \xi, ab\, D(c) - bD(ca)\rangle + \langle \xi, bc\, D(a)\rangle$$

$$= \langle \xi, ab\, D(c) - bD(c)a\rangle$$

$$= \langle \xi a - a\xi, bD(c)\rangle = \langle b \otimes c, F_{\xi a - a\xi}\rangle.$$

Hence, in particular,

$$\langle \xi, af - fa\rangle = \langle \xi a - a\xi, f\rangle = \langle F_{\xi a - a\xi}, \Phi\rangle$$

$$= \langle F_\xi a - aF_\xi, \Phi\rangle + \lim \langle \xi, \bar{\omega}(u_i)D(a)\rangle$$

$$= \lim \langle \xi, \bar{\omega}(u_i)D(a)\rangle = \lim \langle \xi\bar{\omega}(u_i), D(a)\rangle.$$

As X is essential, we obtain

$$\langle \xi, af - fa\rangle = \langle \xi, D(a)\rangle.$$

Therefore $D(a) = af - fa$ and A is amenable.

(2) We assume A to be amenable. By Proposition 1.3, A admits bounded approximate units (u_i). We may determine a subnet (v_s) of (u_i) for which $\lim_s v_s \otimes v_s = u \in (A \hat{\otimes} A)^{**}$ with respect to the weak-*-topology. For every $a \in A$,

$$\pi_{\bar{\omega}}(d_1 u(a)) = \pi_{\bar{\omega}}(au - ua) = 0.$$

Hence we may interpret $d_1 u$ to be a continuous derivation of A into the

Banach A-module $Ker\ tt_{\bar{\omega}}$. As A is amenable, there exists $b \in Ker\ tt_{\bar{\omega}}$ such that $d_1 u = d_1 b$. Let $\Phi = u-b$. For every $a \in A$,

$$a\Phi - \Phi a = d_1\Phi(a) = d_1 u(a) - d_1\ b(a) = 0;$$

moreover,

$$tt_{\bar{\omega}}(\Phi)a = tt_{\bar{\omega}}(u-b)a$$

$$= tt_{\bar{\omega}}(u)a = \lim_s v_s v_s a = a.$$

Hence Φ is a virtual diagonal. □

The virtual diagonal assumes the role performed by the invariant means in the theory of amenable groups.

One immediate criterion for the amenability of a locally compact group relies on an exploitation of the Hahn-Banach theorem in view of extending an invariant functional ([P] 4.D). The next proposition establishes an analogous result for amenable Banach algebras.

<u>PROPOSITION 1.8</u> Let A be a Banach algebra. The following properties are equivalent:

(i) A is amenable.

(ii) For every Banach A-module X and every Banach A-submodule Y, every element of $Z(A,Y^*)$ admits an extension to an element of $Z(A,X^*)$.

(iii) For every Banach A-module X, there exists a projection of X^* onto $Z(A,X^*)$ commuting with any continuous linear operator on X^* that is transposed from a continuous linear operator commuting with the action of A on X.

<u>PROOF</u>: (i) => (ii).

Let $f \in Z(A,Y^*)$. The Hahn-Banach theorem provides an extension $f' \in X^*$ of f. For every $a \in A$, $D(a) = af'-f'a \in Y^{\perp} \subset X^*$; $af'-f'a$ may be considered to be an element of $(X/Y)^*$. So D is a continuous derivation of A into $(X/Y)^*$. As A is amenable, there exists $h \in Y^{\perp} \simeq (X/Y)^*$ such that

$D(a) = ah-ha$ whenever $a \in A$. Therefore $g = f'-h \in Z(A,X^*)$. Moreover, g extends f.

(ii) => (iii).

We equip $X_1 = X^* \hat{\otimes} X$ with the Banach A-module structure defined by

$$a\cdot(f \otimes \xi) = f \otimes a\xi, \quad (f \otimes \xi)a = f \otimes \xi a$$

for $a \in A$, $\xi \in X$, $f \in X^*$. We denote by C the space of continuous linear operators on X commuting with the action of A. Let U [resp. V] be the closed vector subspace of X, generated by $\{{}^tT(f) \otimes \xi - f \otimes T\xi: T \in C, f \in X^*, \xi \in X\}$ [resp. $\{f \otimes \xi: f \in Z(A,X^*), \xi \in X\}$]. Let also W be the closed vector subspace generated by U and V; U and V are Banach A-submodules of W whereas W/U is a Banach A-submodule of X_1/U. We define $\Phi \in X_1{}^*$ by

$$\langle f \otimes \xi, \Phi\rangle = f(\xi),$$

$f \in X^*$, $\xi \in X$. Then $\Phi \in U^\perp \subset X_1{}^*$ as ${}^tT(f)(\xi) = f(T\xi)$ whenever $f \in X^*$, $\xi \in X$, $T \in C$. One may thus define $\Psi \in (X_1/U)^*$ by

$$\Psi(\dot{\eta}) = \Phi(\eta),$$

$\dot{\eta} \in X_1/U$. With respect to the Banach A-module structure considered on X_1, $\Psi \in Z(A,(W/U)^*)$ and, by hypothesis, Ψ admits an extension $\Psi' \in Z(A,(X_1/U)^*)$.

Finally we put

$$\langle \xi, p(f)\rangle = \langle (f \otimes \xi)^\bullet, \psi'\rangle$$

for $f \in X^*$, $\xi \in X$; p is a continuous linear operator of X^* onto $Z(A,X^*)$. If $T \in C$ and $\xi \in X$, $f \in X^*$, we have

$$\langle \xi, p({}^tT(f))\rangle = \langle ({}^tT(f) \otimes \xi)^\bullet, \psi'\rangle$$

$$= \langle (f \otimes T\xi)^\bullet, \psi'\rangle = \langle T\xi, p(f)\rangle = \langle \xi, {}^tTp(f)\rangle;$$

hence p commutes with ${}^tT$ whenever $T \in C$. If $\xi \in X$, $f \in X^*$, we have

$$\langle \xi, p^2(f)\rangle = \langle (p(f) \otimes \xi)^{\cdot}, \Psi'\rangle = \langle p(f) \otimes \xi, \Phi\rangle$$

$$= p(f)(\xi) = \langle \xi, p(f)\rangle;$$

p is a projection onto $Z(A,X^*)$.

(iii) => (i).

If A is nonunital, we put $A_1 = \tilde{A}$; if A is unital, we put $A_1 = A$; let u be the corresponding identity. Let also $X = A_1 \hat{\otimes} A_1$; then $Z(A,X^*) = Z(A_1,X^*)$. For every $a \in A_1$, σ'_a and τ'_a commute with the ordinary action of A_1 on X. By hypothesis, there exists a projection p of X^* onto $Z(A_1,X^*)$ commuting with σ'_a, τ'_a $(a \in A_1)$. We consider $\Phi \in \mathcal{L}(X^*)$ defined by

$$\langle a \otimes b, \Phi(F)\rangle = \langle b \otimes a, F\rangle$$

for $F \in X^*$ and $a,b \in A_1$. Let $M = {}^t\Phi({}^tp(u \otimes u)) \in X^{**}$. We show that M is a virtual diagonal for A_1.

Let $a \in A_1$, $F \in X^*$. Then $\Phi(Fa) = {}^t\sigma'_a(\Phi(F))$ and $\Phi(aF) = {}^t\tau'_a(\Phi(F))$. We have

$$\langle F, aM\rangle = \langle Fa, M\rangle = \langle \Phi(Fa), {}^tp(u \otimes u)\rangle$$

$$= \langle {}^t\sigma'_a(\Phi(F)), {}^tp(u \otimes u)\rangle$$

$$= \langle \sigma'_a(u \otimes u), p(\Phi(F))\rangle$$

$$= \langle u \otimes a, p(\Phi(F))\rangle = \langle (u \otimes u)a, p(\Phi(F))\rangle$$

$$= \langle u \otimes u, ap(\Phi(F))\rangle.$$

Similarly, $\langle F, Ma\rangle = \langle u \otimes u, p(\Phi(F))a\rangle$. Therefore, as $p(\Phi(F)) \in Z(A_1,X^*)$, we have $aM = Ma$.

If $f \in A_1^*$, $a \in A_1$, $b \in A_1$, then ${}^t\bar{\omega}(f) \in X^*$ and, for every $c \in A_1$,

$$\langle a \otimes b, c\Phi({}^t\bar{\omega}(f)) - \Phi({}^t\bar{\omega}(f))c\rangle$$

$$= \langle a \otimes bc - ca \otimes b, \Phi({}^t\bar{\omega}(f))\rangle$$

$$= \langle bc \otimes a - b \otimes ca, {}^t\bar{\omega}(f)\rangle$$

$$= \langle \bar{\omega}(bc \otimes a - b \otimes ca), f\rangle = 0.$$

Hence $\Phi({}^t\bar{\omega}(f)) \in Z(A_1,X^*)$. For all $a \in A_1$, $f \in A_1{}^*$,

$$\langle f, {}^{tt}\bar{\omega}(M)a\rangle = \langle af, {}^{tt}\bar{\omega}(M)\rangle = \langle {}^t\bar{\omega}(af), M\rangle$$

$$= \langle {}^t\bar{\omega}(af), {}^t\Phi({}^tp(u \otimes u))\rangle$$

$$= \langle u \otimes u, p(\Phi({}^t\bar{\omega}(af)))\rangle$$

$$= \langle u \otimes u, \Phi({}^t\bar{\omega}(af))\rangle$$

$$= \langle u \otimes u, {}^t\bar{\omega}(af)\rangle = \langle u, af\rangle = \langle a, f\rangle.$$

We have ${}^{tt}\bar{\omega}(M)a = a$.

By Proposition 1.7, A_1 is amenable. The amenability of A follows from Proposition 0.2. □

B. Stability properties

We establish a collection of stability properties for amenability in the class of Banach algebras.

PROPOSITION 1.9: Let A,B be Banach algebras and let Φ be a continuous homomorphism of A onto a dense subset of B. Then B is also amenable.

PROOF: Let X be a Banach B-module; X constitutes a Banach A-module for the structure defined by $a\xi = \Phi(a)\xi$, $\xi a = \xi\Phi(a)$, $a \in A$, $\xi \in X$. If D is any continuous derivation of B into X*, then $D \circ \Phi$ is a continuous derivation of

A into X^*. By hypothesis, there exists $\eta \in X^*$ such that $D \circ \Phi = d_1\eta$. Hence D coincides with $d_1\eta$ on $\Phi(A)$ and then, by density, also on B. □

We conclude that if A is an amenable Banach algebra and J is a closed ideal, then the quotient algebra A/J is also amenable.

If A is a Banach algebra, we denote by $\Gamma(A)$ the Banach algebra of all *multipliers* of A, i.e., all $T \in \mathcal{L}(A)$ satisfying $T(a)b = T(ab) = aT(b)$ whenever $a, b \in A$. Every $a \in A$ may be identified with an element T of $\Gamma(A)$ if we put $T(x) = ax$ for $x \in A$. More generally, if J is a closed ideal of A, any $a \in A$ may be interpreted as an element of $\Gamma(J)$.

Let A be a Banach algebra admitting bounded approximate units $(u_i)_{i\in I}$ and let X be an essential Banach A-module. If $T \in \Gamma(A)$ and $\xi \in X$, we put

$$T\xi = \lim (Tu_i)\xi, \quad \xi T = \lim \xi(Tu_i);$$

X becomes a Banach $\Gamma(A)$-module. This definition yields the unique extension of the Banach A-module structure to an essential Banach $\Gamma(A)$-module structure. As a matter of fact, if $\xi \in X$ and $T \in \Gamma(A)$, there exist $a, b \in A$ and $\xi, \eta \in X$ such that $\xi = a\eta b$ and therefore

$$T\xi = T(a\eta b) = \lim T(u_i a\eta b) = \lim T(u_i)a\eta b$$
$$= \lim T(u_i)\xi.$$

Thus any continuous derivation of $\Gamma(A)$ into X^* may be restricted to a continuous derivation of A into X^*. Let now D be an inner derivation of A into X^* and let $\xi_0 \in X^*$ such that $D = d_1\xi_0$. For $i \in I$, we put

$$D_i(T) = T(u_i)\xi_0 - \xi_0 T(u_i),$$

$T \in \Gamma(A)$; c being the bound of (u_i), we have $\|D_i\| \leq 2c\|\xi_0\|$ and $D_i \in \mathcal{L}(\Gamma(A), X^*)$. There exists a subnet (u_j') of (u_i) giving rise to a subnet (D_j') of (D_i) such that (D_j') converges to $D_0 \in \mathcal{L}(\Gamma(A), X^*)$ in the $\sigma(X^*, \Gamma(A))$-topology. For every $a \in A$,

$$D_0(a) = \lim (au_j'\xi_0 - \xi_0 au_j') = a\xi_0 - \xi_0 a;$$

$D_o = d_1\xi_o$. We proved that $H_1(\Gamma(A),X^*) \simeq H_1(A,X^*)$.

<u>PROPOSITION 1.10</u>: Let A be a Banach algebra admitting the closed ideal J. If A is amenable and J admits bounded approximate units, then J is also amenable.

<u>PROOF</u>: Let X be an essential Banach J-module; we consider it to be an essential Banach $\Gamma(J)$-module. Thanks to the foregoing statement and Proposition 0.3 we must prove that $H_1(\Gamma(J),X^*) = \{0\}$. Let D be a continuous derivation of $\Gamma(J)$ into X^*. There exists a continuous homomorphism Φ of A into $\Gamma(J)$ such that $D_1 = D\circ\Phi$ is a continuous derivation of A into X^*. We may determine $\xi \in X^*$ such that D_1 coincides with $d_1\xi$ on A; hence D coincides with $d_1\xi$ on $\Gamma(J)$. □

<u>PROPOSITION 1.11</u>: Let A be a Banach algebra and let J be a closed ideal of A. If J and A/J are amenable, then so is A.

<u>PROOF</u>: Let X be a Banach A-module and let D be a continuous derivation of A into X^*. Then $D_1 = D|_J$ is a continuous derivation of J into X^* and, by hypothesis, there exists $\eta \in X^*$ such that $D_1 = d_1\eta$. For all $a \in A$, $b \in J$, we have

$$0 = (D_1-d_1\eta)\ (ab) = (D-d_1\eta)(a)b + a(D-d_1\eta)(b)$$

$$= (D-d_1\eta)(a)b.$$

Thus, for every $\xi \in X$,

$$\langle b\xi,(D-d_1\eta)(a)\rangle = \langle \xi,(D-d_1\eta)(a)b\rangle = 0.$$

Similarly

$$\langle \xi b,(D-d_1\eta)(a)\rangle = 0.$$

The closed subspace X_J of X generated by $JX \cup XJ$ is a Banach A-submodule of X and $D-d_1\eta$ applies A into $X_J^{\perp} = (X/X_J)^*$. Moreover, we may consider

$D_2 = D - d_1\eta$ to be a continuous derivation of A/J into $(X/X_J)^*$. By hypothesis, there exists $\zeta \in X_J^{\perp}$ such that $D_2 = d_1\zeta$. Finally we have $D = d_1(\eta + \zeta)$. □

PROPOSITION 1.12: Let A be a Banach algebra and let $(A_i)_{i\in I}$ be an increasing net of closed subalgebras with $\overline{\bigcup_{i\in I} A_i} = A$. Assume the existence of $k > 0$ such that, for every $i \in I$, every Banach A_i-module X and every continuous derivation D of A_i into X^*, there exists $\eta \in X^*$ satisfying $D = d_1\eta$ and $\|\eta\| \leq k\|D\|$. Then A is amenable.

PROOF: Let X be a Banach A-module and let D be a continuous derivation of A into X^*. For every $i \in I$, choose η_i such that $D|_{A_i} = d_1\eta_i|_{A_i}$ and $\|\eta_i\| \leq k\,\|D\|$. A subnet (ζ_j) of (η_i) is weak-*-convergent to $\eta \in X^*$.

Let $a \in \bigcup_{i\in I} A_i$. There exists $i_o \in I$ such that $a \in A_i$ whenever $i_o < i$. For every $\xi \in X$, we have

$$\langle \xi, D(a)\rangle = \langle \xi, d_1\eta_i(a)\rangle = \langle \xi a - a\xi, \eta_i\rangle.$$

But

$$\lim_j \langle \xi a - a\xi, \zeta_j\rangle = \langle \xi a - a\xi, \eta\rangle = \langle \xi, a\eta - \eta a\rangle.$$

Hence D coincides with $d_1\eta$ on $\bigcup_{i\in I} A_i$ and then, by density, also on A. □

C. Complements

Whereas the amenability of a Banach algebra is defined via the first cohomology group corresponding to a dual Banach module, the next proposition concerns the first cohomology group corresponding to the given Banach module.

PROPOSITION 1.13: Let A be an amenable Banach algebra and let X be a Banach A-module that is abelian, i.e., $a\xi = \xi a$ whenever $a \in A$, $\xi \in X$. Then $H_1(A,X) = \{0\}$.

PROOF: Every continuous derivation D of A into X may be considered to be a continuous derivation of A into $X^{**} = (X^*)^*$; thus it is inner. As X is abelian, we have then $D = 0$. □

We indicate some *examples* of amenable Banach algebras.

PROPOSITION 1.14: If Z is a compact space, $C(Z)$ constitutes an amenable Banach algebra.

PROOF: The abelian group $G = \{\exp ih : h \in C_{\mathbb{R}}(Z)\}$ is amenable ([P] Proposition 12.2); thus $\ell^1(G)$ is an amenable Banach algebra. Consider the homomorphism $f:\ell^1(G) \to C(Z)$ defined by

$$f(\sum_{n \in \mathbb{N}^*} \alpha_n \delta_{\exp ih_n})(z) = \sum_{n \in \mathbb{N}^*} \alpha_n \exp ih_n(z),$$

where (α_n) is a sequence in $\mathbb{C}$ such that $\sum_{n \in \mathbb{N}^*} |\alpha_n| < \infty$, (h_n) is a sequence in $C_{\mathbb{R}}(Z)$, and $z \in Z$; $\| \sum_{n \in \mathbb{N}^*} \alpha_n \exp ih_n \| \leq \sum_{n \in \mathbb{N}^*} |\alpha_n|$, hence $\|f\| \leq 1$, and f is continuous. By Proposition 1.9, $\overline{f(\ell^1(G))}$ is an amenable Banach algebra. The algebra $f(\ell^1(G))$ is self-adjoint for complex conjugation, and there does not exist $z \in Z$ such that $\phi(z) = 0$ whenever $\phi \in f(\ell^1(G))$; one deduces then from the Stone-Weierstrass theorem that $C(Z)$ is an amenable Banach algebra. □

If H is a Hilbert space, we denote by $\mathcal{LC}(H)$ the Banach algebra of *all compact operators* on H; i.e., all $T \in \mathcal{L}(H)$ for which $\overline{T(H_1)}$ is compact, H_1 being the unit ball of H.

PROPOSITION 1.15: Let H be a separable Hilbert space. Then $A = \mathcal{LC}(H) \oplus \mathbb{C}\, id_H$ is an amenable Banach algebra.

PROOF: Let $\{\xi_n : n \in \mathbb{N}^*\}$ be an orthonormal basis of H. Denote by P the group of all permutations p of $\mathbb{N}^*$ such that $p(n) = n$ except for at most a finite number of elements n in $\mathbb{N}^*$. Let Q be the group of all mappings q of $\mathbb{N}^*$ into $\{-1,1\}$ such that $q(n) = 1$ except for at most a finite number of elements n in $\mathbb{N}^*$. If $p \in P$ and $q \in Q$, we define $T = T_{p,q} \in \mathcal{L}(H)$ by putting

$$T(\xi_n) = q(n)\xi_{p(n)},$$

$n \in \underset{\sim}{N}^*$; as $T\text{-}id_H$ has finite rank, $T \in A$. Consider the group $G = \{T_{p,q} : p \in P,\ q \in Q\}$ and the normal subgroup $H = \{T_{id\ \underset{\sim}{N}^*, q} : q \in Q\}$ in G; P and Q, hence also H, are amenable ([P] Proposition 13.6). As $P \simeq G/H$, G is then amenable ([P] Proposition 13.4); the algebra $\ell^1(G)$ is amenable.

The vector space E generated by G is contained in A. In order to prove that $\bar{E} = A$, we show that, if $i,j \in \underset{\sim}{N}^*$, the operator of rank 1 mapping ξ_i on ξ_j and $\xi_{i'}$ $(i' \neq i)$ on 0 belongs to E. Let $p_{i,j}$ be the transposition of the elements i,j in $\underset{\sim}{N}^*$ and, for $n \in \underset{\sim}{N}^*$, let

$$T_1(\xi_n) = \xi_{p_{i,j}(n)},$$

$$T_2(\xi_n) = (1-2\delta_{i,n})\xi_{p_{i,j}(n)},$$

$\delta_{i,n}$ being the Kronecker symbol. Then $\frac{1}{2}(T_1-T_2) \in E$ and $\frac{1}{2}(T_1-T_2)(\xi_i) = \xi_j$, $\frac{1}{2}(T_1-T_2)(\xi_{i'}) = 0$ for $i' \neq i$. We conclude that $A \simeq \ell^1(G)$. □

Comments

We show that a mild perturbation of the multiplicative structure of a Banach algebra associated to a Banach space does not alter the amenable character.

LEMMA 1.16: Let Y and Z be closed subspaces of a Banach space X. Put

$$d = \sup\{\sup\{\inf\{\|\eta-\zeta\| : \zeta \in Z\} : \eta \in Y,\ \|\eta\| = 1\},$$
$$\sup\{\inf\{\|\eta-\zeta\| : \eta \in Y\} : \zeta \in Z,\ \|\zeta\| = 1\}\}.$$

Suppose that $d > 0$ and there exists a projection p of X onto Y such that $\|p\| < d^{-1}-1$. Then $p|_Z$ is a bijection from Z onto Y. For every $\zeta \in Z$,

$$\|\zeta - p\zeta\| \leq d(1 + \|p\|)\ \|\zeta\|.$$

Put $q = (p|_Z)^{-1}$. We have

$$\|q\| \leqq (1+d)\,(1 - \|p\|\,d)^{-1}$$

and, for every $\eta \in Y$,

$$\|\eta - q\eta\| \leqq ((1+d)(1 - \|p\|\,d)^{-1} - 1)\ \|\eta\| \ .$$

PROOF: Let $d_0 \in \,]d, (1 + \|p\|)^{-1}[$. If $\zeta \in Z$, $\zeta \neq 0$, there exists $\eta_\zeta \in Y$ such that $\|\eta_\zeta - \frac{\zeta}{\|\zeta\|}\| \leqq d_0$, hence, for $\eta = \eta_\zeta\ \|\zeta\|$,

$$\|\eta - \zeta\| \leqq d_0\ \|\zeta\| .$$

Trivially, in case $\zeta = 0$, one chooses $\eta = 0$ and the last inequality still holds. Then also

$$\|p\zeta - \eta\| = \|p(\zeta - \eta)\| \leqq \|p\|\ \ \|\zeta - \eta\| \leqq \|p\|\ d_0\ \ \|\zeta\| .$$

If, in particular, $\zeta \neq 0$, then

$$\begin{aligned} \|\zeta - p\zeta\| &\leqq \|\zeta - \eta\| + \|\eta - p\zeta\| \\ &\leqq (1 + \|p\|)d_0\ \|\zeta\| < \|\zeta\| . \end{aligned} \qquad (1)$$

Assume that, for $\zeta, \zeta' \in Z$, $\zeta \neq \zeta'$, one has $p\zeta = p\zeta'$;

$$\|\zeta - \zeta' - p(\zeta - \zeta')\| = \|\zeta - \zeta'\| .$$

We contradict (1). Hence $p|_Z$ is an injection of Z into Y.

Likewise, if $\eta \in Y$, there exists $\zeta \in Z$ such that $\|\zeta - \eta\| \leqq d_0\ \|\eta\|$.

Let us fix $\eta \in Y$ and put $\eta_0 = \eta$. Inductively we define $\eta_1, \eta_2, \ldots \in Y$ and $\zeta_0, \zeta_1, \zeta_2, \ldots \in Z$ such that $\|\zeta_i - \eta_i\| \leqq d_0\ \|\eta_i\|$ for $i \in \underset{\sim}{N}$ and $\eta_i = \eta_{i-1} - p(\zeta_{i-1})$ for $i \in \underset{\sim}{N}^*$. For every $i \in \underset{\sim}{N}^*$, we have

$$\begin{aligned} \|\eta_i\| &= \|p(\eta_{i-1} - \zeta_{i-1})\| \leqq \|p\|\ \ \|\eta_{i-1} - \zeta_{i-1}\| \\ &\leqq \|p\|\,d_0\ \ \|\eta_{i-1}\| ; \end{aligned}$$

thus

$$\|\eta_i\| \leqq \|p\|^i d_o^i \|\eta\|$$

and

$$\|\zeta_i\| \leqq (1+d_o) \|\eta_i\| \leqq (1+d_o) \|p\|^i d_o^i \|\eta\|$$

$$\leqq (1+d_o)\left(\frac{\|p\|}{1+\|p\|}\right)^i \|\eta\| . \qquad (2)$$

Now

$$\zeta' = \sum_{i=0}^{\infty} \zeta_i \in Z \text{ and } p(\zeta') = \sum_{i=0}^{\infty} p(\zeta_i) = \eta_o = \eta.$$

Moreover

$$\|\zeta'\| \leqq \sum_{i=0}^{\infty} \|\zeta_i\| \leqq (1+d_o) \sum_{i=0}^{\infty} \|p\|^i d_o^i \|\eta\|$$

$$= (1+d_o)(1-\|p\| d_o)^{-1} \|\eta\|.$$

We conclude that $p|_Z$ is a bijective continuous linear mapping of Z onto Y and $\|q\| \leqq (1+d_o)(1-\|p\| d_o)^{-1}$; hence, as d_o is chosen arbitrarily in $]d, 1+\|p\|[$,

$$\|q\| \leqq (1+d)(1-\|p\| d)^{-1}.$$

If $\eta \in Y$, by (2),

$$\|\eta-q(\eta)\| = \|\eta-\zeta'\| \leqq \|\eta-\zeta_o\| + \sum_{i=1}^{\infty} \|\zeta_i\|$$

$$\leqq d_o \|\eta\| + (1+d_o) \sum_{i=1}^{\infty} \|p\|^i d_o^i \|\eta\|$$

$$= -\|\eta\| + (1+d_o) \sum_{i=0}^{\infty} \|p\|^i d_o^i \|\eta\|$$

$$= ((1+d_o)(1-\|p\| d_o)^{-1}-1) \|\eta\| .$$

Owing to the possible choices of d_o, we also have

$$\|\eta-q(\eta)\| \leq ((1+d)(1 - \|p\| d)^{-1}-1) \|\eta\| . \qquad \square$$

<u>LEMMA 1.17</u>: Let X_1, X_2, X_3 be Banach spaces. For i = 1,2, consider S_i, $T_i \in \mathcal{L}(X_i,X_{i+1})$ such that $Im\ S_1 = Ker\ S_2$, $Im\ S_2$ is closed and $T_2 \circ T_1 = 0$; let also $k_i \in \underset{\sim}{R}_+^*$ such that, for every $\eta \in Im\ S_i$, there exists $\xi \in X_i$ satisfying $\|\xi\| \leq k_i \|\eta\|$ and $S_i(\xi) = \eta$. Suppose that $k_1 \|S_1-T_1\| + k_2 \|S_2-T_2\| + k_1k_2 \|S_1-T_1\| \|S_2-T_2\| < 1$; then $Im\ T_1 = Ker\ T_2$. In particular, if $S_2 = T_2 = 0$, S_1 is onto and $\|S_1-T_1\| < k_1^{-1}$, then, for every $\eta \in X_2$, there exists $\xi \in X_1$ such that $T_1(\xi) = \eta$ and $\|\xi\| \leq k_1 (1-k_1 \|S_1-T_1\|)^{-1} \|\eta\|$.

<u>PROOF</u>: For i = 1,2, put $\alpha_i = \|S_i-T_i\|$.

Let $\eta \in X_2$; we have

$$\|S_2(\eta)\| \leq \|T_2(\eta)\| + \alpha_2 \|\eta\| .$$

Thus, by hypothesis, there exists $\eta' \in X_2$ such that

$$\|\eta'\| \leq k_2 \|S_2(\eta)\| \leq k_2(\|T_2(\eta)\| + \alpha_2 \|\eta\|) \qquad (3)$$

and

$S_2(\eta') = S_2(\eta)$, i.e., $\eta - \eta' \in Ker\ S_2$. Then, by hypothesis again, there exists $\xi \in X_1$ such that $\|\xi\| \leq k_1 \|\eta-\eta'\|$ and $S_1(\xi) = \eta -\eta'$. We have

$$\begin{aligned} \|\eta-\eta'\| &\leq \|\eta\| + \|\eta'\| \\ &\leq k_2 \|T_2\eta\| + (k_2 \alpha_2 + 1) \|\eta\| ; \end{aligned} \qquad (4)$$

$$\|\xi\| \leq k_1(k_2 \|T_2\eta\| + (k_2 \alpha_2+1) \|\eta\|).$$

Relations (3) and (4) imply that

$$\|\eta - T_1(\xi)\| = \|S_1(\xi) + \eta' - T_1(\xi)\|$$

$$\leqq \|\eta'\| + \|(S_1 - T_1)\xi\|$$

$$\leqq k_2 \|T_2(\eta)\| + k_2\alpha_2 \|\eta\| + \alpha_1 \|\xi\| \qquad (5)$$

$$\leqq (1+k_1\alpha_1) k_2 \|T_2(\eta)\| + (k_1\alpha_1 + k_2\alpha_2 + k_1k_2\alpha_1\alpha_2) \|\eta\| .$$

By hypothesis, $k = k_1\alpha_1 + k_2\alpha_2 + k_1k_2\alpha_1\alpha_2 < 1$.

Let now η be a fixed element in *Ker* T_2. Then (5) reduces to

$$\|\eta - T_1(\xi)\| \leqq k \|\eta\| .$$

By the foregoing procedure we define inductively a sequence (ξ_n) in X_1 and a sequence (η_n) in *Ker* T_2. Let $\xi_o = \xi$, $\eta_o = \eta$. If $\xi_o, \ldots, \xi_n \in X_1$ and $\eta_o, \ldots, \eta_n \in$ *Ker* T_2 have been defined, we put

$$\eta_{n+1} = \eta_n - T_1(\xi_n);$$

then

$$T_2(\eta_{n+1}) = T_2(\eta_n) - T_2 \circ T_1(\xi_n) = T_2(\eta_n) = 0$$

and we may choose $\xi_{n+1} \in X$, in order to have

$$\|\xi_{n+1}\| \leqq k_1(k_2 \|T_2(\eta_{n+1})\| + (k_2\alpha_2+1)\|\eta_{n+1}\|)$$

and

$$\|\eta_{n+1} - T_1(\xi_{n+1})\| \leqq k \|\eta_{n+1}\| .$$

Thus, for every $n \in \underset{\sim}{N}^*$,

$$\|\eta_{n+1}\| = \|\eta_n - T_1(\xi_n)\| \leqq k \|\eta_n\|$$

$$\leqq \ldots \leqq k^{n+1} \|\eta\| \qquad (6)$$

and

$$\|\xi_n\| \leq k_1(k_2 \|T_2(\eta_n)\| + (k_2\alpha_2+1) \|\eta_n\|)$$

$$= k_1(k_2\alpha_2+1) \|\eta_n\| . \tag{7}$$

As $0 \leq k < 1$, we have then $\sum_{n=0}^{\infty} \xi_n = \xi' \in X_1$, and

$$\lim_{n\to\infty} \|\eta - \sum_{j=1}^{n} T_1(\xi_j)\|$$

$$= \lim_{n\to\infty} \|\eta_0 - T_1(\xi_1) \ldots -T_1(\xi_n)\|$$

$$= \lim_{n\to\infty} \|\eta_{n+1}\| \leq \lim_{n\to\infty} k^{n+1} \|\eta\| = 0.$$

Therefore $T_1(\xi') = \eta$ and $\mathit{Ker}\ T_2 \subset \mathit{Im}\ T_1$. As, by hypothesis, $T_2 \circ T_1 = 0$, we finally have $\mathit{Ker}\ T_2 = \mathit{Im}\ T_1$.

In the particular situation considered above, $\alpha_2 = 0$, $k = k_1\alpha_1 < 1$ and, by (7) and (6),

$$\|\xi'\| \leq \sum_{n=0}^{\infty} \|\xi_n\| \leq k_1 \sum_{n=0}^{\infty} \|\eta_n\|$$

$$\leq k_1 \sum_{n=0}^{\infty} k^n \|\eta\| = k_1 (1-k_1\alpha_1)^{-1} \|\eta\| . \quad \square$$

PROPOSITION 1.18: Let A be a Banach space constituting an amenable Banach algebra for the multiplication corresponding to $\bar{\omega}$.

There exists $\delta > 0$ such that, for any multiplication m satisfying $\|m-\bar{\omega}\| < \delta$, (A,m) is an amenable Banach algebra.

PROOF: Let $\varepsilon \in]0,1/4[$ and consider a multiplication defined by $m: A \hat{\otimes} A \to A$ such that $\|m-\bar{\omega}\| < \varepsilon$. We also have

$$\|{}^{tt}m - {}^{tt}\bar{\omega}\| = \|{}^{tt}(m-\bar{\omega})\| < \varepsilon .$$

Thanks to Proposition 0.2, we may suppose the algebra to be unital. Let us consider the mapping

$$\phi: a \mapsto a \otimes u_A.$$

$$A \to A \hat{\otimes} A .$$

We have $\bar{\omega}\circ\phi = id_A$, hence ${}^{tt}\bar{\omega}\circ{}^{tt}\phi = {}^{tt}id_A = id_{A^{**}}$ and $p = id_{(A\hat{\otimes}A)^{**}} - {}^{tt}\phi\circ{}^{tt}\bar{\omega}$ is an element of $\mathcal{L}((A \hat{\otimes} A)^{**}, \mathit{Ker}\ {}^{tt}\bar{\omega})$ with $\|{}^{tt}\phi\| \leqq 1$, $\|p\| \leqq 2$. Moreover,

$$p\circ p = id_{(A\hat{\otimes}A)^{**}} - {}^{tt}\phi\circ{}^{tt}\bar{\omega} - {}^{tt}\phi\circ\ {}^{tt}\bar{\omega} + {}^{tt}\phi\circ\ ({}^{tt}\bar{\omega} \circ {}^{tt}\phi)\circ\ {}^{tt}\bar{\omega}$$

$$= id_{(A\hat{\otimes}A)^{**}} - {}^{tt}\phi\circ\ {}^{tt}\bar{\omega} = p.$$

Let $c \in \mathit{Ker}\ {}^{tt}m$. We have $p(c) \in \mathit{Ker}\ {}^{tt}\bar{\omega}$ and

$$\|c-p(c)\| = \|{}^{tt}\phi\circ\ {}^{tt}\bar{\omega}(c)\| \leqq \|{}^{tt}\bar{\omega}(c)\|$$

$$\leqq \|{}^{tt}m(c)\| + \|{}^{tt}\bar{\omega}(c) - {}^{tt}m(c)\|$$

$$\leqq \varepsilon\|c\| \leqq \varepsilon\ (1-\varepsilon)^{-1}\ \|c\| .$$

We apply the particular case of Lemma 1.17 to

$$X_1 = (A \hat{\otimes} A)^{**},\ X_2 = A^{**},\ S_1 = {}^{tt}\bar{\omega},\ T_1 = {}^{tt}m;$$

$\|S_1-T_1\| < \varepsilon < \frac{1}{4} < 1$. We may assume that $k_1 = 1$. If $a \in \mathit{Ker}\ {}^{tt}\bar{\omega} \subset X_1$, we have ${}^{tt}m(a) \in X_2$; there exists $b \in (A \hat{\otimes} A)^{**}$ such that ${}^{tt}m(b) = {}^{tt}m(a)$ and

$$\|b\| \leqq (1-\varepsilon)^{-1}\|{}^{tt}m(a)\|$$

$$\leqq (1-\varepsilon)^{-1}\ (\|{}^{tt}m(a) - {}^{tt}\bar{\omega}(a)\| + \|{}^{tt}\bar{\omega}(a)\|)$$

$$\leqq (1-\varepsilon)^{-1}\ \varepsilon\ \|a\| .$$

Thus $a-b \in \mathit{Ker}\ {}^{tt}m$ and

$$\|a-(a-b)\| = \|b\| \leqq (1-\varepsilon)^{-1} \varepsilon\|a\| .$$

From the two preceding results we conclude that, for the real constant d associated to the subspaces $\mathit{Ker}\ {}^{tt}\bar{\omega}$, $\mathit{Ker}\ {}^{tt}m$ of $(A \hat{\otimes} A)^{**}$ in Lemma 1.16, we must have

$$d \leqq \varepsilon(1-\varepsilon)^{-1}.$$

As $\varepsilon < 1/4$, also $d < 1/3$ and $d^{-1} -1 > 2 \geqq \|p\|$. We now apply Lemma 1.16. There exists a surjection $q \in \mathcal{L}(\mathit{Ker}\ {}^{tt}\bar{\omega}, \mathit{Ker}\ {}^{tt}m)$ such that $\|q^{-1}\| \leqq 2$ and

$$\|q\| \leqq (1+d)(1 - \|p\|\, d)^{-1} \leqq (1 + \frac{\varepsilon}{1-\varepsilon})(1 - \frac{\varepsilon\|p\|}{1 - \varepsilon})^{-1}$$

$$= (1-(\|p\| +1)\varepsilon)^{-1} < (1 - 3\varepsilon)^{-1}.$$

For every $a \in \mathit{Ker}\ {}^{tt}m$,

$$\|a-q^{-1}(a)\| \leqq d(1 + \|p\|) \|a\| \leqq 3\varepsilon(1-\varepsilon)^{-1} \|a\| . \tag{8}$$

For every $b \in \mathit{Ker}\ {}^{tt}\bar{\omega}$,

$$\|b-q(b)\| \leqq ((1+d)(1 - \|p\|\, d)^{-1}-1) \|b\| \tag{9}$$

$$\leqq ((1-3\varepsilon)^{-1}-1) \|b\| = 3\varepsilon(1-3\varepsilon)^{-1} \|b\| .$$

On $(A \hat{\otimes} A)^{**}$ we consider the Banach module structures associated to the Banach algebras defined by $\bar{\omega}$ and m. For all $x \in A$ and $F \in (A \hat{\otimes} A)^{**}$,

$$\|x\cdot_{\bar{\omega}}F-x\cdot_{m}F\| \leqq \varepsilon\|x\| \|F\| ,$$

$$\|F\cdot_{\bar{\omega}}x-F\cdot_{m}x\| \leqq \varepsilon \|x\| \|F\| . \tag{10}$$

We also consider d_1, d_2 for the Banach module $\mathit{Ker}\ {}^{tt}\bar{\omega}$ with respect to $(A,\bar{\omega})$ and the corresponding mappings $d_{1,m}, d_{2,m}$ for the Banach module $\mathit{Ker}\ {}^{tt}m$ with respect to (A,m). For i = 1,2, we put

$$d_i' = q^{-1} \circ d_{i,m} \circ q.$$

For all $x \in A$ and $a \in Ker\ ^{tt}_{\bar{\omega}}$, by (10), (9), (8) we obtain

$$\begin{aligned} &\|x\cdot_{\bar{\omega}} a - q^{-1}(x \cdot_m q(a))\| \\ &\leq \|x\cdot_{\bar{\omega}} a - x\cdot_{\bar{\omega}} q(a)\| + \|x\cdot_{\bar{\omega}} q(a) - x\cdot_m q(a)\| \\ &\quad + \|x\cdot_m q(a) - q^{-1}(x\cdot_m q(a))\| \\ &\leq O(\varepsilon)\, \|x\|\ \|a\|. \end{aligned}$$

Therefore,

$$\|d_1 - d_1'\| \leq O(\varepsilon)$$

and

$$\|d_2 - d_2'\| \leq O(\varepsilon).$$

As $(A,\bar{\omega})$ is amenable, we have $Im\ d_1 = Ker\ d_2$; moreover, $Im\ d_2$ is closed. If $\xi \in Im\ d_1'$, there exists $a \in Ker\ ^{tt}_{\bar{\omega}}$ such that $\xi = d_1'(a) = q^{-1}\circ d_{1,m}\circ q(a)$. Then

$$d_2'(\xi) = q^{-1}\circ d_{2,m}\circ q\circ q^{-1}\circ d_{1,m}\circ q(a) = q^{-1}\circ d_{2,m}\circ d_{1,m}\circ q(a).$$

As $Im\ d_{1,m} \subset Ker\ d_{2,m}$, we have thus $d_2'(\xi) = 0$. Hence $d_2'\circ d_1' = 0$. We now apply Lemma 1.17 to d_1, d_2, d_1', d_2'; we may conclude that $Im\ d_1' = Ker\ d_2'$.

If $a \in Ker\ d_{2,m}$, we have $d_2'\circ q^{-1}(a) = q^{-1}\circ d_{2,m}(a) = 0$; hence $q^{-1}(Ker\ d_{2,m}) \subset Ker\ d_2'$. If $b \in q(Im\ d_1')$, there exists $c \in Ker\ ^{tt}_{\bar{\omega}}$ such that $b = q\circ d_1'(c) = d_{1,m}\circ q(c)$; hence $q(Im\ d_1') \subset Im\ d_{1,m}$. Therefore, $Ker\ d_{2,m} \subset q(Ker\ d_2') = q(Im\ d_1') \subset Im\ d_{1,m}$. As also $Im\ d_{1,m} \subset Ker\ d_{2,m}$, we conclude that

$$Ker\ d_{2,m} = Im\ d_{1,m}.$$

We define a continuous derivation D of A into $Ker\ {}^{tt}m$ by putting

$$D(a) = a \otimes u_A - u_A \otimes a,$$

for $a \in A$. There exists then $\xi \in Ker\ {}^{tt}m$ such that $D = d_{1,m}\xi$. Define also $\Phi = u_A \otimes u_A - \xi \in (A \hat{\otimes} A)^{**}$.

Let $a \in A$. We have $a._m(u_A \otimes u_A) - (u_A \otimes u_A)._m a = D(a) = a._m\xi - \xi._m a$, hence $a._m\Phi = \Phi._m a$. Moreover,

$${}^{tt}m(\Phi)._m a = {}^{tt}m(u_A \otimes u_A - \xi)._m a = {}^{tt}m(u_A \otimes u_A)._m a = a.$$

Therefore Φ is a virtual diagonal for (A,m). Amenability of the latter algebra now follows from Proposition 1.7. □

The class of amenable algebras is stable by passage to tensor products.

PROPOSITION 1.19: If A and B are amenable Banach algebras, so is $A \hat{\otimes} B$.

PROOF: As A and B admit bounded approximate units by Proposition 1.3, so does $A \hat{\otimes} B$. With respect to Proposition 0.3, it suffices to prove that, for any essential Banach $A \hat{\otimes} B$-module X, $H_1(A \hat{\otimes} B, X^*) = \{0\}$.

We may extend $\Gamma(A) \hat{\otimes} \Gamma(B)$ injectively into $\Gamma(A \hat{\otimes} B)$. The mappings $a \to \sigma_a$, $b \to \tau_b$ determine continuous homomorphisms of A, B onto closed subalgebras A_1, B_1 of $\Gamma(A \hat{\otimes} B)$; A_1, B_1 commute and, by Proposition 1.9, these algebras are amenable. Moreover, X may be considered to be a Banach A_1-module and a Banach B_1-module.

Let $D \in H_1(A \hat{\otimes} B, X^*)$; it gives rise to an element D' of $H_1(\Gamma(A \hat{\otimes} B), X^*)$. We denote by D_1 the restriction of D' to B_1. As B_1 is amenable, there exists $\eta \in X^*$ such that $D_1 = d_1\eta$. Put $D'' = D' - d_1\eta$; D" is a continuous derivation of $\Gamma(A \hat{\otimes} B)$ into X* vanishing on B_1. For all $a \in A$ and $b \in B$, we have

$$\tau_b\, D''(\sigma_a) = D''(\tau_b\sigma_a) = D''(\sigma_a\tau_b) = D''(\sigma_a)\tau_b,$$

hence $\tau_b\, D''(\sigma_a) - D''(\sigma_a)\tau_b = 0$ and $\{D''(\sigma_a) : a \in A\}$ is contained in $Y = (X/Im\ d)^*$, where d denotes the mapping d_1 defined for B_1. As A_1 and B_1

commute, *Im* d is a Banach A_1-module. Therefore, Y is the dual of the Banach A_1-module X/*Im* d. Moreover, the restriction of D" to A_1 is a continuous derivation of A_1 into Y. As A_1 is amenable, there exists $\zeta \in Y$ such that $D'' = d_1\zeta$ on A_1. As $\zeta \in Y$, $D'' = 0 = d_1\zeta$ on B_1. Then $D''' = D'' - d_1\zeta$ vanishes on $A_1 \cup B_1$. Hence the derivation D is inner on $A \hat{\otimes} B$. □

Notes

Johnson initiates the theory of amenable Banach algebras [42]. An introductory exposition may be found in [44]. Johnson proves 1.2-3 [42]. The notions of approximate diagonal and virtual diagonal are also defined by that author; he establishes 1.6-7 [43]. Bunce [6] proves the equivalence (i) <=> (iii) in 1.8 for unital C*-algebras; the general demonstration and the other implications are due to Lau [53]. Khelemskiǐ and Sheinberg [50] produce a modified version of the fundamental definition of amenable Banach algebras. See also Racher [60].

Johnson proves the stability properties 1.9-12 [42]. Johnson also observes 1.13 [42]. The examples 1.14-15 are elaborated by Bonsall and Duncan [5].

Stegmeir [64] produces refinements for the amenability properties of Banach algebras. He proves that any maximal modular ideal of an amenable Banach algebra admits bounded approximate units.

For other examples of amenable Banach algebras we refer to Johnson [42]; he shows that the Banach algebra of compact linear operators on $C(\underset{\sim}{R}/\underset{\sim}{Z})$ is amenable. For a locally compact group G, Stegmeir [64] examines amenability of Banach subalgebras of $L^1(G)$ consisting of functions that remain invariant under the actions of subgroups in the automorphism group of G.

Let A be a Banach algebra and let X be a Banach A-module. For $n = 2,3,\ldots$, one defines a continuous linear mapping of the Banach space, determined by the tensor product of n copies of A and X, into the Banach space, determined by the tensor product of n-1 copies of A and X; if $a_1,\ldots,a_n \in A$ and $\xi \in X$, one puts

$$d_n(a_1 \otimes \ldots \otimes a_n \otimes \xi) = a_2 \otimes \ldots \otimes a_n \otimes \xi a_1$$

$$+ \sum_{i=1}^{n-1} (-1)^i \, a_1 \otimes \ldots \otimes a_i a_{i+1} \otimes \ldots \otimes a_n \xi$$

$$+ (-1)^n \, a_1 \otimes \ldots \otimes a_{n-1} \otimes a_n \xi.$$

Then $d_n \circ d_{n+1} = 0$. The n-th cohomology group is $H_n(A,X) = \mathit{Ker}\, d_n / \mathit{Im}\, d_{n+1}$. We quote some of Johnson's results concerning amenability in relation with these cohomology groups of higher orders. If A is an abelian Banach algebra and X is an abelian dual Banach A-module, then $H_n(A,X) = \{0\}$ for every $n \in \underset{\sim}{N}^*$ [40]. Let A be an infinite-dimensional semisimple Banach algebra. If the algebra is abelian, there exists a Banach A-module X such that $H_1(A,X) \neq \{0\}$ [42]; if the algebra is amenable, there exists a Banach A-module X such that $H_2(A,X) \neq \{0\}$ [43]. Johnson produces also an amenable Banach algebra A and a Banach A-module X satisfying $H_n(A,X) = \{0\}$ whenever $n \geqq 2$ [43].

2 Amenable C*-algebras

First we summarize some supplementary facts about C*-algebras.

Let A be a C*-algebra. Any mapping $\theta: A_+ \to \overline{\mathbb{R}_+}$ is called *trace* on A_+ if

(a) $\theta(x + y) = \theta(x) + \theta(y)$ whenever $x, y \in A_+$,

(b) $\theta(\alpha x) = \alpha\theta(x)$ whenever $\alpha \in \mathbb{R}_+$, $x \in A_+$,

(c) $\theta(aa^*) = \theta(a^*a)$ whenever $a \in A$.

The trace is said to be *faithful* if $\theta(x) > 0$ for every $x \in A_+$ such that $x \neq 0$. The trace is called *finite* in case it admits its values in $\mathbb{R}_+$; it is called *semifinite* if, for every $x \in A_+$, $\theta(x)$ is the least upper bound of the numbers $\theta(y)$, where $y \in A_+$, $y \leqq x$ and $\theta(y) < \infty$. Finally we say the trace is *normal* if, for every increasing net (x_i) in A_+ admitting the least upper bound x, we have $\theta(x) = \sup_i \theta(x_i)$. To the trace θ we associate the ideal n_θ formed by all $x \in A$ such that $\theta(xx^*) < \infty$. On $m_\theta = n_\theta^2$ there exists a unique functional θ', coinciding with θ on $m_{\theta,+}$; for all $x, y \in n_\theta$, $\theta'(xy) = \theta'(yx)$ ([D'] 6.1.2).

Let A be a C*-algebra. The *representation* π of A admitting the Hilbert space $H_\pi = H$ as its representation space is a homomorphism $\pi: x \mapsto \pi_x$; $A \to \mathcal{L}(H)$ it is automatically continuous. If $x \in A$ and $\xi, \eta \in H_\pi$, the number $(\pi_x\xi|\eta) = (\pi(x)\xi|\eta)$, called coefficient, is also denoted by $\omega_{H;\pi;\xi,\eta}(x) = \omega_{\xi,\eta}(x)$; we write ω_ξ for $\omega_{\xi,\xi}$. The representation is called **-representation* if $\pi_x{}^* = \pi_{x^*}$ whenever $x \in A$. Two representations π, π', admitting the representation spaces H_π, $H_{\pi'}$, are termed *equivalent* if there exists an isometry U of H_π onto $H_{\pi'}$ such that $U^{-1}\pi'_x U = \pi_x$ whenever $x \in A$. Every vector $\xi \in H_\pi$ for which $\{\pi_x\xi : x \in A\}^- = H$ is said to be *cyclic* with respect to the representation π of A.

If H is a direct sum $\bigoplus_{i \in I} H_i$ of Hilbert spaces and, for every $i \in I$, $\pi^{(i)}$ is a *-representation of A on H_i, we define a *-representation π of A on H by putting

$$\pi_x \xi = \sum_{i \in I} \pi_x^{(i)} \xi_i,$$

$$\pi_x^* \xi = \sum_{i \in I} \pi_x^{(i)*} \xi_i,$$

for $\xi = \sum_{i \in I} \xi_i \in H$, $x \in A$ ([T] p.41). We write $\pi = \sum_{i \in I} \pi^{(i)}$.

Any linear mapping of an involutive algebra A into an involutive algebra B is called *positive* if it transforms any element of the form a*a (a ∈ A) into an element of the form b*b (b ∈ B). We term *state* on the C*-algebra A every (continuous) positive linear functional f on A that is normalized, i.e., $\|f\| = 1$; in case A is unital, the latter condition signifies that f(u) = 1, u being the unit of A. We denote by $S(A)$ the positive cone of all states on A.

Let H be a Hilbert space. We say that $T \in \mathcal{L}(H)$ is a *positive-definite* operator if $(T\xi|\xi) \geq 0$ whenever $\xi \in H$; then $T = T^*$. For every $T \in \mathcal{L}(H)$, T*T is positive-definite. The positive cone of positive-definite operators on H coincides with $\mathcal{L}(H)_+$ ([D'] 1.6.7).

Let A be a C*-algebra. If π is a *-representation of A and $\xi \in H_\pi$, then

$$f : x \mapsto (\pi_x \xi | \xi)$$
$$A \to \mathbb{C}$$

is a positive functional on A. Conversely, if f is a positive functional on A, there exists a *-representation π of A on a Hilbert space H and $\xi \in H$ such that $f(x) = (\pi_x \xi | \xi)$ whenever $x \in A$; this representation is unique up to an equivalence ([T] Chapter I. Theorem 9.14).

Let A be a commutative unital C*-algebra. There exists a compact set Ω, the spectrum of A, such that A is isomorphic to $C(\Omega)$ via a mapping $\chi : A \to C(\Omega)$ satisfying $\chi(x^*) = \overline{\chi(x)}$ for every $x \in A$. If $x \in A$, $\chi(x^*x) \geq 0$; there exists $y = (x^*x)^{\frac{1}{2}} \in A_+$ such that $\chi(y) \geq 0$ and $y^2 = x^*x$.

Let H be a Hilbert space and let T be a positive-definite operator in $\mathcal{L}(H)$. Considering the abelian C*-subalgebra A of $\mathcal{L}(H)$, generated by T and id_H, and the corresponding spectrum Ω, we may identify A with $C(\Omega)$. There exists a positive element U in A such that $U^2 = T$; U is the square root of T and is denoted by $T^{\frac{1}{2}}$. Thanks to the Weierstrass theorem, U corresponds, in $C(\Omega)$, to a limit of polynomials over the variable T; hence, if T commutes with a closed subalgebra B of $\mathcal{L}(H)$, so does U.

If H is a Hilbert space, every $T \in \mathcal{L}(H)$ admits a unique *polar decomposition* $T = W|T|$, where $|T| = (T^*T)^{\frac{1}{2}}$ and W is a partial isometry.

The spectral theory, applied to the unital abelian C*-algebra generated by a hermitian element T of $\mathcal{L}(H)$ for a Hilbert space H, enables also the determination of a 1-parameter family $\{P_t : t \in \mathbb{R}\}$ of *spectral projections* satisfying the following conditions: a) If $t, t' \in \mathbb{R}$ and $t \leq t'$, one has $P_t P_{t'} = P_t$; b) if $t \in \mathbb{R}$ and $\xi \in H$, one has $\lim_{\varepsilon \downarrow 0} \|P_{t-\varepsilon}\xi - P_t\xi\| = 0$; if $t \in \mathbb{R}$, P_t commutes with every operator in $\mathcal{L}(H)$ commuting with T ([D] Appendice I.3).

Let A be a unital C*-algebra. We consider the involutive algebra $\mathfrak{M}_n(A)$ of all $n \times n$ matrices $a = [a_{ij}]$ with entries in A, the involution being defined by $(a^*)_{ij} = a_{ji}^*$ for $a = [a_{ij}]$; $i,j = 1,\ldots,n$. The matrix $a = [a_{ij}] \in \mathfrak{M}_n(A)$ may be interpreted as an operator T_a on $\mathbb{C}^n$ if one puts

$$T_a(z_1,\ldots,z_n) = \left(\sum_{j=1}^{n} a_{1j}z_j, \ldots, \sum_{j=1}^{n} a_{nj}z_j\right)$$

for $(z_1,\ldots,z_n) \in \mathbb{C}^n$.

Let A and B be C*-algebras. If ϕ is a linear mapping of A into B and $n \in \mathbb{N}^*$, one considers the linear mapping

$$\phi_n : \mathfrak{M}_n(A) \to \mathfrak{M}_n(B)$$

defined by $\phi_n(a) = [\phi(a_{ij})]$ for $a = [a_{ij}] \in \mathfrak{M}_n(A)$. If ϕ_n is positive, ϕ is called n-*positive*; this situation is realized if and only if $\sum_{i,j=1}^{n} y_i^* \phi(x_i^* x_j) y_j \geqq 0$ whenever $x_1,\ldots,x_n \in A$ and $y_1,\ldots,y_n \in B$ ([T] Chapter IV. Corollary 3.4). In case ϕ is n-positive for every $n \in \mathbb{N}^*$, it is said to be *completely positive*.

Let A be a C*-algebra and let f be a positive functional on A. One considers an associated representation π on a Hilbert space H and a cyclic vector ξ in H. One may define a continuous linear mapping θ_f of $\widetilde{\pi(A)}$ into the Banach space that is generated by the positive cone of all elements $g \in A^*$ for which there exist $\alpha = \alpha_g > 0$ such that $0 \leq g \leq \alpha f$; if

$a \in A$ and $x \in \pi(A)^{\sim}$,

$$\langle a, \theta_f(x)\rangle = (\pi_a x\xi | \xi).$$

The mapping θ_f is bijective and completely positive; the inverse mapping is also completely positive ([T] Chapter IV. Proposition 3.10). If ϕ is a completely positive mapping of the unital C*-algebra B into A* such that $\phi(u_B) = f$, then the mapping $\theta_f^{-1} \circ \phi : B \to \pi(A)^{\sim}$ is completely positive ([T] Chapter IV. Remark 3.11 (i)).

Let A [resp. B] be a C*-algebra admitting the unit u_A [resp. u_B]. We call *morphism* of A into B every completely positive linear mapping ϕ of A into B such that $\phi(u_A) = u_B$. Let $f \in (A \otimes_\gamma B)^*$. Then $f \in S(A \otimes_\gamma B)$ if and only if Γ_f is a completely positive mapping of B into A* and $\Gamma_f(u_B)$ is a state on A ([T] Chapter IV. Proposition 4.6). More generally, every $F:B \to A^*$ that is completely positive and for which $F(u_B)$ is a state constitutes a *completely normalized* mapping.

Let A_1 and A_2 be C*-algebras. The tensor product $A_1 \otimes A_2$ is equipped with an involution by putting

$$(x_1 \otimes x_2)^* = x_1^* \otimes x_2^*$$

for $x_1 \in A_1$, $x_2 \in A_2$. The Banach algebra $A_1 \otimes_\gamma A_2$ is involutive. The norm $\|\cdot\|$ on $A_1 \otimes A_2$ is called *C*-norm* if

$$\|xy\| \leqq \|x\| \ \|y\|$$

whenever $x,y \in A_1 \otimes A_2$, and

$$\|x^*x\| = \|x\|^2$$

whenever $x \in A_1 \otimes A_2$.

Let Σ be the set of all involutive representations of the algebraic tensor product $A_1 \otimes A_2$, i.e., all homomorphisms $\pi : A_1 \otimes A_2 \to \mathcal{L}(H)$, for a Hilbert space H, such that $\pi(x^*) = \pi(x)^*$ whenever $x \in A_1 \otimes A_2$. If $x \in A_1 \otimes A_2$, one puts

$$\|x\|_{max} = \sup \{\|\pi(x)\| : \pi \in \Sigma\};$$

$\|\cdot\|_{max}$ is a C*-norm called projective norm. The completion of $A_1 \otimes A_2$ for $\|\cdot\|_{max}$ is a C*-algebra denoted by $A_1 \otimes_{max} A_2$.

If π_1 is a *-representation of A_1 on the Hilbert space H_1 and π_2 is a *-representation of A_2 on the Hilbert space H_2, there exists $\pi_1 \otimes \pi_2 \in \Sigma$, $\pi_1 \otimes \pi_2 : A_1 \otimes A_2 \to \mathcal{L}(H_1 \otimes H_2)$ such that

$$\pi_1 \otimes \pi_2(x) = \sum_{i=1}^{n} \pi_1(x_{1,i}) \otimes \pi_2(x_{2,i})$$

whenever $x = \sum_{i=1}^{n} x_{1,i} \otimes x_{2,i} \in A_1 \otimes A_2$.

If $x \in A_1 \otimes A_2$, one denotes by $\|x\|_{min}$ the least upper bound of the numbers $\|\pi(x)\|$, where $\pi = \pi_1 \otimes \pi_2$ and π_1 [resp. π_2] runs over the set of all *-representations of A_1 [resp. A_2]; $\|\cdot\|_{min}$ is a C*-norm called injective norm. The completion of $A_1 \otimes A_2$ for $\|\cdot\|_{min}$ is a C*-algebra denoted by $A_1 \otimes_{min} A_2$.

$\|\cdot\|_{min}$ is the smallest C*-norm; $\|\cdot\|_{min} \leqq \|\cdot\|_{max}$. Both these norms are cross-norms ([T] Chapter IV. Theorem 4.19).

If A and B are C*-algebras, and A_o is a C*-subalgebra of A, then the canonical injection of $A_o \otimes_{min} B$ into $A \otimes_{min} B$ is an isometry ([T] Chapter IV. Proposition 4.22).

Let A and B be C*-algebras. We still denote by $S(A \otimes B)$ the set of all states on $A \otimes B$, i.e., all functionals f on $A \otimes B$, of norm 1, for which $f(x^*x) \geqq 0$ whenever $x \in A \otimes B$. Every $f \in S(A \otimes B)$ may be extended continuously to an element $f' \in S(A \otimes_{max} B)$. If $f \in S(A \otimes B)$ and $x \in A \otimes B$, we put

$$p_f(x) = f(x^*x)^{1/2};$$

p_f is a seminorm on $A \otimes B$ and $p_f(x^*x) = p_f(x)^2$ whenever $x \in A \otimes B$ ([D'] 2.7.1). If Γ is a nonvoid subset in $S(A \otimes B)$, $p_\Gamma = \sup \{p_f : f \in \Gamma\}$ constitutes also a seminorm on $A \otimes B$. In case p_Γ is a norm, i.e., for every $x \in A \otimes B$, $p_\Gamma(x) = 0$ if and only if $x = 0$, we say that Γ is a *separating* family in $S(A \otimes B)$.

We consider the separating families max = $S(A \otimes B)$ and min = $S(A \otimes B) \cap (A^* \otimes B^*)$; $p_{max} = \|\cdot\|_{max}$, $p_{min} = \|\cdot\|_{min}$. The corresponding enveloping C*-algebras are $A \otimes_{max} B$ and $A \otimes_{min} B$ ([D'] 2.7.1).

In general, if Γ is a separating family in $S(A \otimes B)$, we denote by $A \otimes_{\Gamma} B$ the enveloping C*-algebra determined by Γ. We also put $S_{\Gamma}(A \otimes B) = \{f|_{A \otimes B} : f \in S(A \otimes_{\Gamma} B)\}$. Hence, we may identify $S_{max}(A \otimes B)$ with $S(A \otimes B)$. If Γ_1, Γ_2 are separating families in $S(A \otimes B)$ for which $\Gamma_1 \subset \Gamma_2$, then $p_{\Gamma_1} \leq p_{\Gamma_2}$ and there exists a canonical homomorphism of $A \otimes_{\Gamma_2} B$ onto $A \otimes_{\Gamma_1} (B)$.

If A and B are C*-algebras, the associated norms $\|\cdot\|_{max^*}$ and $\|\cdot\|_{min^*}$ coincide on $A^* \otimes B^*$ ([T] Chapter IV. Proposition 4.10). But in general, $A \otimes_{max} B$ and $A \otimes_{min} B$ are different; hence the following definition is of importance.

DEFINITION 2.1: The C*-algebra A is nuclear if, for every C*-algebra B, $A \otimes_{max} B = A \otimes_{min} B$.

Let Φ denote the canonical homomorphism of $A \otimes_{max} B$ onto $A \otimes_{min} B$. Every state on $A \otimes_{min} B$ may be identified with a state on $A \otimes_{max} B$ that vanishes on $\mathit{Ker}\ \Phi$; A is nuclear if, for every C*-algebra B, every state on $A \otimes_{max} B$ is of that type.

If A and B are C*-algebras, any C*-norm on $A \otimes B$ may be extended to a C*-norm on $\tilde{A} \otimes \tilde{B}$ ([T] Chapter IV. Corollary 4.3). Hence, in the study of norms on tensor products of C*-algebras, the latter may be assumed to be unital.

PROPOSITION 2.2: The C*-algebra A is nuclear if and only if, for every unital C*-algebra B, every completely normalized mapping of B into A* may be approximated, in the weak operator topology, by completely normalized mappings of B into A* having finite rank.

PROOF: (1) Suppose A to be nuclear.

Let H, K be Hilbert spaces on which A, B operate faithfully, i.e., injectively; $A \otimes_{max} B = A \otimes_{min} B$ operates faithfully on $H \otimes K$. Then

$S(A \otimes_{max} B)$ is the weak-*-closed convex hull of the set of all states f defined by unitary vectors $\sum_{j=1}^{n} \xi_j \otimes \eta_j$, where $\xi_1,\ldots,\xi_n \in H$, $\eta_1,\ldots,\eta_n \in K$, $n \in \mathbb{N}^*$, i.e., for $a_1,\ldots,a_m \in A$ and $b_1,\ldots,b_m \in B$,

$$f\left(\sum_{i=1}^{m} a_i \otimes b_i\right) = \left(\left(\sum_{i=1}^{m} a_i \otimes b_i\right)\left(\sum_{j=1}^{m} \xi_j \otimes \eta_j\right) \middle| \sum_{k=1}^{n} \xi_k \otimes \eta_k\right)$$

([D'] 2.4.3, 3.4.2). For all $a \in A$, $b \in B$,

$$\Gamma_f(b)(a) = \sum_{j=1}^{n} \sum_{k=1}^{n} (a\xi_j \otimes b\eta_j | \xi_k \otimes \eta_k)$$

$$= \sum_{j=1}^{n} \sum_{k=1}^{n} (a\xi_j|\xi_k)_H \, (b\eta_j|\eta_k)_K$$

$$= \sum_{j=1}^{n} \sum_{k=1}^{n} \omega_{\xi_j,\xi_k}(a) \; \omega_{\eta_j,\eta_k}(b),$$

i.e., $\Gamma_f(b) = \sum_{j=1}^{n} \sum_{k=1}^{n} \omega_{\eta_j,\eta_k}(b)\, \omega_{\xi_j,\xi_k}$. Hence, Γ_f has finite rank.

One obtains the required approximation property.

(2) Conversely, let $f \in S(A \otimes_{max} B)$ for which Γ_f has finite rank; there exist $f_1,\ldots,f_n \in A^*$ and $g_1,\ldots,g_n \in B^*$ such that $\Gamma_f(b) = \sum_{j=1}^{n} f_j(a)g_j(b)$ whenever $a \in A$ and $b \in B$. Hence $f = \sum_{j=1}^{n} f_j \otimes g_j$ and, for $x \in A \otimes B$,

$$|f(x)| \leq \sum_{j=1}^{n} \|f_j\| \, \|g_j\| \, \|x\|_{min};$$

the mapping f is continuous on $A \otimes_{min} B$ and vanishes on the kernel of the canonical homomorphism Φ mapping $A \otimes_{max} B$ onto $A \otimes_{min} B$. Hence $S(A \otimes_{max} B)$ is the set of all states defined on $A \otimes_{max} B$ and vanishing on *Ker* Φ. □

A. Unital amenable C*-algebras

We formulate characterizations for amenability of unital C*-algebras.

LEMMA 2.3: Let A be a C*-algebra with unit u. Assume the existence of a continuous linear mapping Φ of $(A \hat{\otimes} A)^*$ into $Z(A,(A \hat{\otimes} A)^*)$ such that $\Phi(a\cdot f) = a\cdot\Phi(f)$, $\Phi(f\cdot a) = \Phi(f)\cdot a$ whenever $f \in (A \hat{\otimes} A)^*$ and $a \in A$. Let X be a Banach A-module and let D be a continuous derivation of A into X*. If $\xi \in X$, one defines $f_\xi \in (A \hat{\otimes} A)^*$ by putting

$$f_\xi(b \otimes c) = \langle \xi c, D(b)\rangle$$

for $b,c \in A$. If $\xi \in X$ and $a \in A$, one defines $g_{\xi,a} \in (A \hat{\otimes} A)^*$ by putting

$$g_{\xi,a}(b \otimes c) = \langle \xi cb, D(a)\rangle$$

for $b,c \in A$. Finally, one defines $F \in X^*$ by letting

$$F(\xi) = \langle u \otimes u, \Phi(f_\xi)\rangle$$

for $\xi \in X$. Then, $g_{\xi,a} \in Z(A,(A \hat{\otimes} A)^*)$ and $d_1F(a)(\xi) = \Phi(g_{\xi,a})$ $(u \otimes u)$ whenever $a \in A$ and $\xi \in X$.

PROOF: For all $d \in A$ and $b,c \in A$,

$$d\, g_{\xi,a}(b \otimes c) = g_{\xi,a}((b \otimes c)d) = g_{\xi,a}(b \otimes cd)$$

$$= D(a)(\xi(cd)b) = D(a)\ (\xi c(db)) = g_{\xi,a}(db \otimes c)$$

$$= g_{\xi,a}(d(b \otimes c)) = g_{\xi,a}d(b \otimes c);$$

therefore $g_{\xi,a} \in Z(A,(A \hat{\otimes} A)^*)$.

If $a,b \in A$ and $\xi \in X$, we have

$$(f_\xi.a)(b \otimes c) = f_\xi(b \otimes ac) = D(b)(\xi ac)$$

and

$$f_{\xi a}(b \otimes c) = D(b)(\xi_{ac});$$

hence

$$f_\xi \cdot a = f_{\xi a}. \tag{1}$$

Moreover,

$$a \cdot f_\xi(b \otimes c) = f_\xi(ba \otimes c) = \langle \xi c, D(ba)\rangle$$

$$= \langle \xi c, D(b)a\rangle + \langle \xi c, bD(a)\rangle$$

$$= \langle a\xi c, D(b)\rangle + \langle \xi cb, D(a)\rangle$$

$$= f_{a\xi}(b \otimes c) + g_{\xi,a}(b \otimes c);$$

hence

$$a \cdot f_\xi = f_{a\xi} + g_{\xi,a}. \tag{2}$$

For $a \in A$ and $\xi \in X$, from (1), (2) and the fact that $\Phi(f_\xi) \in Z(A,(A \hat{\otimes} A)^*)$ we obtain

$$d_1F(a)(\xi) = (aF-Fa)(\xi) = F(\xi a - a\xi) = \Phi(f_{\xi a - a\xi})(u \otimes u)$$

$$= \Phi(f_\xi \cdot a - a \cdot f_\xi + g_{\xi,a})(u \otimes u)$$

$$= \Phi(f_\xi)(u \otimes a - a \otimes u) + \Phi(g_{\xi,a})(u \otimes u)$$

$$= (a\Phi(f_\xi) - \Phi(f_\xi)a)(u \otimes u) + \Phi(g_{\xi,a})(u \otimes u)$$

$$= \Phi(g_{\xi,a})(u \otimes u). \quad \square$$

PROPOSITION 2.4: Let A be a unital C*-algebra. The following properties are equivalent:

(i) A is amenable.

(ii) If X is a Banach A-module, Y is a Banach A-submodule of X, and $F \in Y^*$ such that $F(v\xi v^*) = F(\xi)$ whenever $\xi \in Y$, $v \in A_u$, then there exists $F_1 \in X^*$ extending F such that $F_1(v\xi v^*) = F_1(\xi)$ whenever $\xi \in X$, $v \in A_u$.

(iii) There exists a continuous linear operator Φ of $(A \hat{\otimes} A)^*$ into $Z(A,(A \hat{\otimes} A)^*)$ such that $\Phi(f) = f$ whenever $f \in Z(A,(A \hat{\otimes} A)^*)$, and $\Phi(a\cdot f) = a\cdot\Phi(f)$, $\Phi(f\cdot a) = \Phi(f)\cdot a$ whenever $f \in (A \hat{\otimes} A)^*$ and $a \in A$.

PROOF:

(i) <=> (ii) is given by the equivalence (i) <=> (ii) in Proposition 1.8.

The implications (i) => (iii) is a particular case of the implication (i) => (iii) of Proposition 1.8.

(iii) => (i)

With the notations of Lemma 2.3, for all $a \in A$, $\xi \in X$, we have

$$\Phi(g_{\xi,a}) = g_{\xi,a}$$

and

$$d_1F(a)(\xi) = g_{\xi,a}(u \otimes u) = D(a)(\xi). \qquad \square$$

We add an extension property for amenability on C*-algebras.

PROPOSITION 2.5: Let A be a C*-algebra admitting the unit u, generated by an amenable involutive Banach subalgebra B and an element a in A such that $a \in B$, $\|a\| \leq 1$, $a^*a = u$ and $axa^* \in B$ whenever $x \in B$. Then A is amenable.

PROOF: (i) Let $e = aa^* = aua^* \in B$. Then $e^2 = aa^*aa^* = aua^* = aa^* = e$ and $ea = (aa^*)a = a(a^*a) = a$. For every $n \in \underset{\sim}{N}^*$,

$$a^{*n}\, a^n = u^n = u.$$

By recurrence, we see that, for every $n \in \underset{\sim}{N}^*$,

$$a^n \quad a^{*n} = a(a^{n-1}xa^{*(n-1)})a^* \in B$$

whenever $x \in B$.

(2) We consider a unital Banach A-module X and a continuous derivation D of A into X*. As B is amenable, there exists $\xi \in X^*$ such that $D|_B = d_1\xi$. Let $D_1 = D - d_1$. We have to show that D_1 is inner.

Notice that $D_1(e) = D_1(aa^*) = 0$ and

$$D_1(a)a^* + aD_1(a^*) = D_1(aa^*) = 0;$$

hence

$$D_1(a^*) = uD_1(a^*) = a^*aD_1(a^*) = -a^*D_1(a)a^*. \tag{3}$$

(3) If $n \in \underset{\sim}{N}^*$, put

$$\zeta_n = a^{*n}D_1(a^n) = D_1(a^{*n}a^n) - D_1(a^{*n})a^n = -D_1(a^{*n})a^n;$$

$$\|\zeta_n\| \leqq \|D_1\|.$$

Let ζ be a weak-*-limit point of (ζ_n) in X*.

For all $n \in \underset{\sim}{N}^*$ and $x \in B$,

$$\begin{aligned} x\zeta_n - \zeta_n x &= xa^{*n}D_1(a^n) - a^{*n}D_1(a^n)x \\ &= a^{*n}a^nxa^{*n}D_1(a^n) - a^{*n}D_1(a^n)x \\ &= a^{*n}((a^nxa^{*n})D_1(a^n) - D_1(a^n)x). \end{aligned}$$

But

$$\begin{aligned} D_1(a^nx) = D_1(a^nxa^{*n}a^n) &= (a^nxa^{*n})D_1(a^n) + D_1(a^nxa^{*n})a^n \\ &= a^nxa^{*n}D_1(a^n). \end{aligned}$$

We obtain then

$$x\zeta_n - \zeta_n x = a^{*n}(D_1(a^n x) - D_1(a^n)x)$$

$$= a^{*n}\ a^n D_1(x) = 0.$$

Therefore

$$x\zeta - \zeta x = 0. \tag{4}$$

Moreover, for every $n \in \underset{\sim}{N}^*$,

$$eD_1(a) = eD_1(a^{*n}a^n a) = e(D_1(a^{*n})a^{n+1} + a^{*n}D_1(a^{n+1}))$$

$$= e(D_1(a^{*n})a^n a + aa^{*(n+1)}D_1(a^{n+1}))$$

$$= e(-\zeta_n a + a\zeta_{n+1}).$$

Therefore

$$eD_1(a) = e(-\zeta a + a\zeta).$$

With respect to (4) we obtain

$$eD_1(a) = -\zeta ea + ea\zeta = -\zeta a + a\zeta.$$

As X is unital,

$$0 = D_1(a-a) = D_1((u-e)a)$$

$$= (u-e)D_1(a) + D_1(u-e)a$$

$$= (u-e)D_1(a) = D_1(a) - eD_1(a).$$

Thus

$$D_1(a) = a\zeta - \zeta a. \qquad (5)$$

From (3) and (5) we deduce that

$$D_1(a^*) = -a^*(a\zeta - \zeta a)a^* = a^*\zeta e - \zeta a^*.$$

From (4) we obtain then

$$D_1(a^*) = a^* e\zeta - \zeta a^* = a^*aa^*\zeta - \zeta a^*,$$

$$D_1(a^*) = a^*\zeta - \zeta a^*. \qquad (6)$$

By (4), (5), (6) the restriction of D_1 to $B \cup \{a,a^*\}$ coincides with $d_1\zeta$. Hence D_1 is inner on A. □

B. Strongly amenable C*-algebras

Let G be a locally compact group and let X be a unital Banach G-module. One may define a new unital Banach G-module structure on X; put $x\cdot\xi = \xi$ and $\xi\cdot x = x^{-1}\xi x$ for $x \in G$ and $\xi \in X$. To a continuous derivation D of A into X* for the given structure one may associate the continuous derivation $E: x \mapsto D(x)x^{-1}$ for the new structure. The locally compact group $G \to X^*$
G is amenable if and only if, for every Banach G-module structure on $L^1(G)$ such that $xf = f$ whenever $x \in G$ and $f \in L^1(G)$, and every continuous derivation D of G into $L^\infty(G) = L^1(G)^*$, there exists an element Φ in the closed convex hull of $\{D(x) : x \in G\}$ such that $D = -d_1\Phi$ ([P] Lemma 11.7. Proposition 11.9).

The following definition may be considered to be a transposition of that situation.

DEFINITION 2.6: The C*-algebra A is called strongly amenable if, for every Banach A-module X and every continuous derivation D of A into X*, there exists $\eta \in \overline{\mathrm{co}}\,\{D_1(x)x^* : x \in \tilde{A}_u\}$ such that $D = -d_1\eta$, D_1 being the extension of D to $\tilde{A}$.

Obviously, any strongly amenable C*-algebra is amenable.

In the particular case of strongly amenable C*-algebras, the properties of Proposition 2.4 admit more precise versions.

LEMMA 2.7: Let A be a unital Banach algebra and let X be a Banach A-module. One considers a nonvoid set Ω of elements T in $\mathcal{L}(X^*)$ for which $T(f) \in \overline{co}\ \{vfv^* : v \in A_u\}$ whenever $f \in X^*$. If $(T_i)_{i \in I}$ is a totally ordered net in Ω, there exists $T_o \in \mathcal{L}(X^*)$ such that $T_o(f)(\xi) = \lim_i T_i(f)(\xi)$ whenever $f \in X^*$, $\xi \in X$, and $T_o(f) \in \overline{co}\{vT_i(f)v^* : v \in A_u\}$ whenever $f \in X^*$, $i \in I$.

PROOF: The set Ω belongs to the closed unit ball of $\mathcal{L}(X^*) \simeq (X \hat{\otimes} X^*)^*$ which is weak-*-compact. For any $i \in I$, consider the weak-*-closure S_i of $\{T_j : i < j\}$. Then $\bigcap_{i \in I} S_i \neq \phi$; let $T_o \in \bigcap_{i \in I} S_i$. For all $f \in X^*$, $\xi \in X$, we have

$$T_o(f)(\xi) = \lim_i T_i(f)(\xi).$$

Let $i \in I$ and $f \in X^*$. Consider $K = \overline{co}\ \{vT_i(f)\ v^* : v \in A_u\}$. If $T_o(f)$ does not belong to K, then, by the Hahn-Banach theorem, there exist $\xi \in X$, $\alpha \in \mathbb{R}$, $\varepsilon \in \mathbb{R}_+^*$ such that, for every $g \in K$,

$$\mathcal{Re}\ T_o(f)(\xi) \leqq \alpha < \alpha + \varepsilon \leqq \mathcal{Re}\ g(\xi);$$

hence, in particular,

$$\alpha + \varepsilon \leqq \mathcal{Re}\ T_i(f)(\xi).$$

We come to a contradiction.

So we conclude that $T_o(f) \in K$. □

PROPOSITION 2.8: Let A be a unital C*-algebra. The following properties are equivalent:

(i) A is strongly amenable.

(ii) For every Banach A-module X and any $f \in X^*$, there exists

$g \in \overline{co}\{vfv^* : v \in A_u\}$ such that $ag = ga$ whenever $a \in A$.

(iii) For every $f \in (A \hat{\otimes} A)^*)$ there exists $g \in \overline{co}\{vfv^* : v \in A_u\}$ such that $ag = ga$ whenever $a \in A$.

(iv) There exists a continuous linear mapping Φ of $(A \hat{\otimes} A)^*$ into $Z(A,(A \hat{\otimes} A)^*)$ such that $\Phi(a \cdot f) = a \cdot \Phi(f)$, $\Phi(f \cdot a) = \Phi(f) \cdot a$ and $\Phi(f) \in \overline{co}\{vfv^* : v \in A_u\}$ whenever $a \in A$, $f \in (A \hat{\otimes} A)^*$.

(v) If X is a Banach A-module and S is a weak-*-closed convex subset of X^* such that $vsv^* \in S$ whenever $v \in A_u$ and $s \in S$, then there exists $t \in S$ such that $vtv^* = t$ whenever $v \in A_u$.

(vi) If X is a Banach A-module, Y is a closed subspace of X such that $v\eta v^* \in Y$ whenever $\eta \in Y$, $v \in A_u$, and $f \in Y^*$ such that $f(v\eta v^*) = f(\eta)$ for all $\eta \in Y$, $v \in A_u$, then, for every $g \in X^*$ extending f, there exists $h \in \overline{co}\{vgv^* : v \in A_u\}$ extending f such that $h(v\xi v^*) = h(\xi)$ whenever $\xi \in X$, $v \in A_u$.

(vii) If X is a Banach A-module, Y is a Banach A-submodule of X, and $f \in Y^*$ such that $f(v\eta v^*) = f(\eta)$ for all $\eta \in Y$, $v \in A_u$, then for every $g \in X^*$ extending f, there exists $h \in \overline{co}\{vgv^* : v \in A_u\}$ extending f such that $h(v\xi v^*) = h(\xi)$ whenever $\xi \in X$, $v \in A_u$.

Proof: We denote by u the unit of A.

(i) => (ii)

There exists $h \in \overline{co}\{d_1f(v)v^* : v \in A_u\}$ such that $d_1f = -d_1h$. For every $v \in A_u$, $d_1f(v)v^* = vfv^* - f$; hence $g = f + h \in \overline{co}\{vfv^* : v \in A_u\}$. Moreover, for every $a \in A$, $d_1f(a) + d_1h(a) = 0$, i.e., $ag = ga$.

(ii) => (iii)

Trivial

(iii) => (iv)

Let Ω denote the set of all continuous endomorphisms T on $(A \hat{\otimes} A)^*$ such that $T(f) \in \overline{co}\{vfv^* : v \in A_u\}$ and $T(a \cdot f) = a \cdot T(f)$, $T(f \cdot a) = T(f) \cdot a$ whenever $f \in (A \hat{\otimes} A)^*$, $a \in A$. We have $id_{(A \hat{\otimes} A)^*} \in \Omega$. We introduce an order relation for Ω. If $T', T'' \in \Omega$, we put $T' < T''$ in case $\overline{co}\{vT''(f)v^* : v \in A_u\} \subset \overline{co}\{vT'(f)v^* : v \in A_u\}$ whenever $f \in (A \hat{\otimes} A)^*$. Let

$\{T_i : i \in I\}$ be a totally ordered subfamily in Ω. By Lemma 2.7 there exists $T_0 \in \mathcal{L}((A \hat{\otimes} A)^*)$ such that

$$T_0(f)(x) = \lim_i T_i(f)(x)$$

whenever $f \in (A \hat{\otimes} A)^*$, $x \in A \hat{\otimes} A$, and

$$T_0(a \cdot f) = a \cdot T_0(f), \; T_0(f \cdot a) = T_0(f) \cdot a$$

whenever $f \in (A \hat{\otimes} A)^*$, $a \in A$. Moreover, for all $f \in (A \hat{\otimes} A)^*$ and $i \in I$,

$$T_0(f) \in \overline{co} \{vT_i(f)v^* : v \in A_u\}$$

and

$$\overline{co}\{vT_0(f)v^* : v \in A_u\} \subset \overline{co}\{vT_i(f)v^* : v \in A_u\}.$$

Hence $T_0 \in \Omega$ and $T_i < T_0$ for every $i \in I$. By Zorn's lemma, the family $\{T_i : i \in I\}$ admits a maximal element U.

Suppose that there exists $h \in (A \hat{\otimes} A)^*$ such that $U(h) \notin Z(A,(A \hat{\otimes} A)^*)$. Then $K = \overline{co}\{vU(h)v^* : v \in A_u\}$ does not reduce to $\{U(h)\}$. By hypothesis, $Z(A,(A \hat{\otimes} A)^*) \cap \overline{co}\{vU(h)v^* : v \in A_u\} \neq \emptyset$. In the closed convex hull of the operators

$$f \mapsto vU(f)v^*$$

in $\mathcal{L}((A \otimes A)^*)$, where $v \in A_u$, consider a family (U_p) for which $(U_p(h))$ converges to $k \in Z(A,(A \hat{\otimes} A)^*)$. By Lemma 2.7 there exists $V \in \mathcal{L}((A \hat{\otimes} A)^*)$ such that

$$V(f)(x) = \lim_p U_p(f)(x)$$

whenever $f \in (A \hat{\otimes} A)^*$, $x \in A \hat{\otimes} A$, and

$$V(f) \in \overline{co}\{vU(f)v^* : v \in A_u\} \tag{7}$$

whenever $f \in (A \hat{\otimes} A)^*$.

For all $b,c \in A$ and $v \in A_u$,

$$vU(a\cdot f)v^*(b \otimes c) = v\,U(a\cdot f)(v^*b \otimes c)$$

$$= U(a\cdot f)(v^*b \otimes cv) = a\cdot U(f)\,(v^*b \otimes cv)$$

$$= U(f)((v^*b \otimes cv)\cdot a) = U(f)\,(v^*ba \otimes cv)$$

$$= vU(f)v^*(ba \otimes c) = vU(f)v^*((b \otimes c)\cdot a).$$

Therefore, we have also

$V(a\cdot f)(b \otimes c) = V(f)((b \otimes c)\cdot a) = a\cdot V(f)(b \otimes c)$. Hence, $V(a\cdot f) = a\cdot V(f)$; similarly, $V(f\cdot a) = V(f)\cdot a$. Now $V \in \Omega$, and by (7), $U < V$. But $V(h) = k \in Z(A,(A \hat{\otimes} A)^*)$. Thus

$$\overline{co}\{vV(h)v^*: v \in A_u\} = \{k\} = \{V(h)\} \subsetneq \overline{co}\{vU(h)v^*: v \in A_u\},$$

$U \neq V$. As U is maximal, a contradiction arises. So we conclude that $U(h) \in Z(A,(A \hat{\otimes} A)^*)$.

(iv) => (v)

Fix $s \in S$. For $\xi \in X$, we define $F_\xi \in (A \hat{\otimes} A)^*$ putting

$$F_\xi(a \otimes b) = s(a\xi b),$$

$a,b \in A$. Let also

$$F(\xi) = \Phi(F_\xi)\,(u \otimes u),$$

$\xi \in X$; $F \in X^*$. For all $v \in A_u$ and $a,b \in A$,

$$F_{v^*\xi v}(a \otimes b) = s(av^*\xi vb) = F_\xi(av^* \otimes vb)$$

$$= F_\xi(v\cdot(a \otimes b)\cdot v^*) = v^*.F_\xi\cdot v(a \otimes b);$$

hence $F_{v^*\xi v} = v^*.F_\xi\cdot v$. As Φ is a mapping into $Z(A,(A \hat{\otimes} A)^*)$, for all $\xi \in X$ and $v \in A_u$,

$$vFv^*(\xi) = F(v^*\xi v) = \Phi(F_{v^*\xi v})(u \otimes u)$$

$$= \Phi(v^*.F_\xi\cdot v)(u \otimes u) = v\cdot\Phi(F_\xi)\cdot v^*(u \otimes u)$$

$$= \Phi(F_\xi)(v^*\cdot(u \otimes u)\cdot v) = \Phi(F_\xi)(v \otimes v^*)$$

$$= v^*\Phi(F_\xi)v\ (u \otimes u) = \Phi(F_\xi)v^*v(u \otimes u)$$

$$= \Phi(F_\xi)(u \otimes u) = F(\xi).$$

Therefore $vFv^* = F$ whenever $v \in A_u$.

If F did not belong to S, there would exist $\xi \in X$, $\alpha \in \mathbb{R}$, $\varepsilon \in \mathbb{R}_+^*$ such that

$$\mathcal{Re}\ F(\xi) \leqq \alpha < \alpha + \varepsilon < \mathcal{Re}\ t(\xi)$$

whenever $t \in S$. But $\Phi(F_\xi) \in \overline{\text{co}}\{vF_\xi v^* : v \in A_u\}$ and, for every $v \in A_u$,

$$vF_\xi v^*(u \otimes u) = F_\xi(v^* \otimes v) = s(v^*\xi v) = vsv^*(\xi).$$

As $vsv^* \in S$ whenever $v \in A_u$, we should then have

$$\mathcal{Re}\ F(\xi) \geqq \alpha + \varepsilon.$$

We conclude that $F \in S$.

(v) = (vi)

Consider $S = \overline{\text{co}}\{vgv^*: v \in A_u\}$ in the Banach A-module X^*. If $s \in S$, then $wsw^* \in S$ for every $w \in A_u$. By hypothesis, there exists $h \in S$ such that $vhv^* = h$ whenever $v \in A_u$; hence $h(v\xi v^*) = h(\xi)$ whenever $\xi \in X$ and $v \in A_u$. As $vfv^* = f$ whenever $v \in A_u$, and g extends f, we conclude that h extends f.

(vi) => (vii)

Trivial

(vii) => (iii)

Let $X = A \otimes A$, $Y = \{0\}$, $f = 0$, and $g \in (A \hat{\otimes} A)^*$. By hypothesis, there exists $h \in \overline{co}\,\{vgv^* : v \in A_u\}$ such that $h(v\xi v^*) = h(\xi)$ whenever $\xi \in X$, $v \in A_u$. Hence, for every $v \in A_u$, $v^*hv = h$ and $hv = vh$.

(iv) => (i)

We adopt again the notations of Lemma 2.3.

For all $a \in A$, $v \in A_u$, $\xi \in X$, we have

$$vg_{\xi,a}v^*(u \otimes u) = g_{\xi,a}(v^* \otimes v)$$

$$= \langle \xi vv^*, D(a)\rangle = \langle \xi, D(a)\rangle :$$

moreover, for all $a \in A$, $\xi \in X$,

$$\Phi(g_{\xi,a}) \in \overline{co}\,\{vg_{\xi,a}v^* : v \in A_u\}$$

and

$$\langle \xi, d_1F(a)\rangle = \langle \xi, D(a)\rangle.$$

If $v \in A_u$ and $\xi \in X$, we have

$$vf_\xi v^*(u \otimes u) = f_\xi(v^* \otimes v) = \langle \xi v, D(v^*)\rangle$$

$$= \langle \xi, vD(v^*)\rangle = \langle \xi, D(vv^*) - D(v)v^*\rangle$$

$$= \langle \xi, -D(v)v^*\rangle.$$

For every $\xi \in X$, $\Phi(f_\xi) \in \overline{co}\{vf_\xi v^* : v \in A_u\}$; therefore $F(\xi)$ belongs to the closed convex hull of $\{\langle \xi, -D(v)v^*\rangle : v \in A_u\}$. If we had $-F \notin \overline{co}\{D(v)v^* : v \in A_u\}$, then, by the Hahn-Banach theorem, there would exist $\alpha \in \underset{\sim}{R}$, $\varepsilon \in \underset{\sim+}{R}^*$, $\xi \in X$ such that

$$-Re\ F(\xi) \leqq \alpha < \alpha + \varepsilon \leqq Re\ \ T(\xi)$$

whenever $T \in \overline{co}\{D(v)v^* : v \in A_u\}$; a contradiction would arise. Thus we must have $-F \in \overline{co}\{D(v)v^* : v \in A_u\}$. □

We indicate a stability property in the class of strongly amenable C*-algebras.

PROPOSITION 2.9: Let A be a C*-algebra and let $(A_i)_{i\in I}$ be an increasing net of strongly amenable C*-subalgebras for which $\overline{\bigcup_{i\in I} A_i} = A$. Then A is strongly amenable.

PROOF: For every $i \in I$, $(\tilde{A}_i)_u$ may be considered to be a subgroup of $\tilde{A}_u$. Let X be a Banach A-module and let D be a continuous derivation of A into X*. For every $i \in I$, there exists

$$\eta_i \in \overline{co}\{D(x)x^* : x \in (\tilde{A}_i)_u\} \subset \overline{co}\{D(x)x^* : x \in \tilde{A}_u\}$$

such that $D|_{A_i} = -d\eta_i$; (η_i) admits a weak-*-limit point η in $\overline{co}\{D(x)x^* : x \in \tilde{A}_u\}$.

Let $i \in I$, $x \in A_i$, $\xi \in X$ and $\varepsilon \in \underset{\sim+}{R}^*$. There exists $j \in I$ such that $A_j \supset A_i$ and

$$|\langle\xi,(D + d_1\eta)(x)\rangle| = |\langle\xi,(-d_1\eta_j + d_1\eta)(x)|$$

$$= |\langle x\xi-\xi x,\ \eta_j-\eta\rangle| < \varepsilon.$$

As $\varepsilon \in \underset{\sim+}{R}^*$ may be chosen arbitrarily, $D|_{A_i} = -d_1\eta|_{A_i}$ for every $i \in I$; hence $D = -d_1\eta$ on A. □

PROPOSITION 2.10: Every unital, strongly amenable C*-algebra A admits a finite trace.

PROOF: We denote by u the unit of A and consider A to be a Banach module on itself.

Let f be a state on A. For all $a \in A$ and $v \in A_u$,

$$\langle a^*a, vfv^*\rangle = \langle v^*a^*av, f\rangle = \langle (av)^*(av), f\rangle \geqq 0;$$

for every $v \in A_u$,

$$\langle u, vfv^*\rangle = \langle v^*v, f\rangle = \langle u, f\rangle = 1.$$

By Proposition 2.8, there exists $g \in \overline{co}\{vfv^* : v \in A_u\}$ such that $ag = ga$ whenever $a \in A$.

We have $\langle u,g\rangle = 1$. For every $a \in A$,

$$0 \leqq \langle a^*a, g\rangle < \infty$$

and

$$\langle a^*a, g\rangle = \langle a, ga^*\rangle = \langle u, aga^*\rangle$$

$$= \langle a^*, ag\rangle = \langle a^*, ga\rangle = \langle aa^*, g\rangle.$$

Hence g is a finite trace on A. □

<u>Comments</u>

We denote by $L^2(G,H)$ the Hilbert space formed by the square-integrable functions defined on the locally compact group G with values in H; we put

$$\|f\|_2 = \left(\int_G \|f(s)\|^2 ds\right)^{\frac{1}{2}}$$

for $f \in L^2(G,H)$. There exists an isometry Φ of $H \otimes L^2(G)$ onto $L^2(G,H)$ defined by

$$\Phi(\xi \otimes g)(s) = g(s)\xi$$

where $\xi \in H$, $g: G \to \underset{\sim}{C}$ is a continuous bounded function with compact support, and $s \in G$ ([69] Part I. Proposition 2.2). We identify $L^2(G,H)$ with $H \otimes L^2(G)$.

Let A be a C*-algebra, considered to be a C*-subalgebra of $\mathcal{L}(H)$ for a separable Hilbert space H, and let G be a locally compact group for which A is a Banach G-module. Let $f \in L^2(G,H)$. If $x \in A$, consider

$$\lambda_x f(s) = sxs^{-1}f(s),$$

$s \in G$; if $t \in G$, consider

$$\lambda'_t f(s) = f(t^- s),$$

$s \in G$. The C*-algebra generated by $\{\lambda_x : x \in A\} \cup \{\lambda'_t : t \in G\}$ in $\mathcal{L}(L^2(G,H))$ is called C*-*crossed product* of G and A; it is denoted by C*(G,A).

PROPOSITION 2.11: Let G be an amenable discrete group and let A be an amenable C*-algebra constituting a Banach G-module. Then the crossed product C*(G,A) is amenable.

PROOF: We extend the action of G over $\tilde{A}$; let $B = C^*(G,\tilde{A})$. We may identify $\tilde{A}$ with a Banach subalgebra of B and consider G to be a subgroup of B_u.

Let X be a Banach B-module and let $D: B \to X^*$ be a continuous derivation; X constitutes also a Banach $\tilde{A}$-module. As A is amenable, so is $\tilde{A}$ by Proposition 0.2; $D|_{\tilde{A}}$ is inner, i.e., there exists $F \in X^*$ such that $D(a) = aF-Fa$ whenever $a \in \tilde{A}$. Let $D_1(a) = aF-Fa$ for $a \in B$; $D-D_1$ is a continuous derivation with $(D-D_1)|_{\tilde{A}} = 0$. Hence we may suppose that $D|_{\tilde{A}} = 0$.

We put $f(x) = xD(x^{-1})$ for $x \in G$; let M be a left invariant mean on $\ell^\infty(G)$. If $x \in G$ and $\xi \in X$, we put

$$f_\xi(x) = \langle \xi, f(x) \rangle;$$

we define $f' \in X^*$ by

$$\langle \xi, f' \rangle = M(f_\xi),$$

$\xi \in X$.

Let $a \in \tilde{A}$ and $x \in G$. As $D|_{\tilde{A}} = 0$, we have $D(x^{-1}ax) = 0$ and

$$af(x) - f(x)a = axD(x^{-1}) - xD(x^{-1})a$$

$$= x(x^{-1}axD(x^{-1}) - D(x^{-1})a)$$

$$= x(D(x^{-1}a) - D(x^{-1})a) = D(a) = 0.$$

Hence, for every $\xi \in X$,

$$0 = \langle \xi, af(x) - f(x)a \rangle = \langle \xi a - a\xi, f(x) \rangle$$

and

$$0 = \langle \xi a - a\xi, f' \rangle = \langle \xi, af' - f'a \rangle,$$

i.e.,

$$af' - f'a = 0 = D(a).$$

If $x, y \in G$,

$$yf(x) = yxD(x^{-1}) = yxD((yx)^{-1}y)$$

$$= yx(D((yx)^{-1})y + (yx)^{-1}D(y))$$

$$= f(yx)y + D(y),$$

$$D(y) = yf(x) - f(yx)y.$$

But, for every $\xi \in X$,

$$\langle \xi, yf(x) - f(yx)y \rangle = \langle \xi y, f(x) \rangle - \langle y\xi, f(yx) \rangle$$

$$= f_{\xi y}(x) - f_{y\xi}(yx).$$

As M is a left invariant mean, we have then

$$\langle \xi, D(y)\rangle = M(f_{\xi y}) - M(f_{y\xi})$$

$$= \langle \xi y, f'\rangle - \langle y\xi, f'\rangle = \langle \xi, yf'-f'y\rangle;$$

$$D(y) = yf'-f'y.$$

As $\tilde{A}$ and G generate $C^*(G,\tilde{A})$, we obtain $D = d_1 f'$. We conclude that $B = C^*(G,\tilde{A})$ is amenable; the unital C^*-algebra $C^*(G,A)$ being a closed ideal in $C^*(G,\tilde{A})$, we deduce amenability of $C^*(G,A)$ from Proposition 1.10. □

Let A be a unital C^*-algebra generated by two isometries U_1, U_2 over a Hilbert space H such that $U_1U_1^* + U_2U_2^* = id_H$, $U_1^*U_1 = U_2^*U_2 = id_H$. The algebra satisfies the conditions of Proposition 2.5 ([62]); it is amenable. If the algebra A were also strongly amenable, by Proposition 2.10 it would admit a finite trace θ such that

$$1 = \theta(id_H) = \theta(U_1U_1^* + U_2U_2^*)$$

$$= \theta(U_1U_1^*) + \theta(U_2U_2^*) = \theta(U_1^*U_1) + \theta(U_2^*U_2),$$

whereas

$$\theta(U_1^*U_1) = \theta(U_2^*U_2) = \theta(id_H) = 1.$$

Thus there exists an amenable C^*-algebra that is not strongly amenable.

In Proposition 3.1 we show that not every C^*-algebra is amenable.

Notes

The concept of nuclearity is due to Takesaki [66]; he calls it property [T]. The term itself is introduced by Lance [52] who proves 2.2.

Bunce [6] [8] establishes 2.3 and gives a complete demonstration of 2.4. Rosenberg [61] proves 2.5.

The notion of strong amenability is introduced by Johnson [42]. He demonstrates 2.9 and a collection of other stability properties. Let A be a C^*-algebra admitting a closed ideal J. If A is (strongly) amenable, so is A/J. If A/J and J are strongly amenable, so is A. Let A and B be strongly amenable C*-algebras; then $A \otimes_\beta B$, for any cross-norm β, is strongly amenable. Every C*-algebra admits a maximal strongly amenable ideal. Johnson also provides several examples of strongly amenable C*-algebras [42]. The demonstration of 2.7 goes back to J.T. Schwartz [62]. Bunce [6] proves 2.8. Rosenberg [61] establishes 2.10.

Rosenberg [61] proves 2.11 and extends the result to strongly amenable C*-algebras.

The example of the amenable C*-algebra that is not strongly amenable is studied by Cuntz [21] in a general context; he shows, in particular, that this algebra satisfies condition 2.5 which Rosenberg [61] proves to imply amenability. Paschke [59] considers a unital strongly amenable C*-algebra A operating on a Hilbert space H admitting $T \in \mathcal{L}(H)$ such that a) $T^*T = id_H$, b) $TT^* \neq id_H$, c) $TAT^* \subset A$, d) $T^*AT \subset A$, e) any closed ideal J of A for which $TJT^* \subset J$ is A or $\{0\}$. He proves that the C*-algebra generated by A and T in $\mathcal{L}(H)$ is simple, i.e., admits no nontrivial closed ideal.

Takesaki [66] shows that every abelian C*-algebra as well as $\mathcal{L}(H)$, over a separable Hilbert space H, are nuclear. The nuclearity of a C*-algebra is studied in detail by Effros and Lance [23]. Other results related to these considerations are due to Kirchberg [51], Kadison and Ringrose[48].

Choi and Effros [12] prove that, given a C*-algebra A and a closed ideal J, A is nuclear if and only if J and A/J are; if A is nuclear C*-algebra and B is a separable C*-subalgebra there exists a separable nuclear C*-algebra between B and A. Nuclear C*-algebras may admit nonnuclear C*-subalgebras [15].Blackader [4] gives a demonstration, due to Effros, establishing the existence of a separable C*-algebra that is not isomorphic to a subalgebra of a nuclear C*-algebra. In this context see also Zillmann [72].

At the recent Luxembourg Harmonic Analysis Conference (September, 1987), A.Paterson announced remarkable results on locally compact groups G for which the corresponding C*-algebras C*(G) are amenable; the class of these groups containsall amenable groups, all type I groups, all almost connected groups.

3 Amenability and von Neumann algebras

Let H be a Hilbert space. For every $x \in \mathcal{L}(H)$,

$$\|x\| = \sup \{|(x\xi|\xi)| : \xi \in H, \|\xi\| = 1\}.$$

In $\mathcal{L}(H)$ consider the a) *strong*, b) *weak*, c) *ultrastrong*, d) *ultraweak topologies* defined by the seminorms

a) $T \mapsto \|T\xi\|$; $\xi \in H$, b) $T \mapsto |(T\xi|\eta)|$; $\xi, \eta \in H$,

c) $T \mapsto (\sum_{i=1}^{\infty} \|T\xi_i\|^2)^{\frac{1}{2}}$; $\xi_1, \xi_2, \ldots \in H$, $\sum_{i=1}^{\infty} \|\xi_i\|^2 < \infty$,

d) $T \mapsto |\sum_{i=1}^{\infty} (T\xi_i|\eta_i)|$; $\xi_1, \xi_2, \ldots \in H$, $\eta_1, \eta_2, \ldots \in H$, $\sum_{i=1}^{\infty} \|\xi_i\|^2 < \infty$,

$\sum_{i=1}^{\infty} \|\eta_i\|^2 < \infty$. For every involutive subalgebra in $\mathcal{L}(H)$, the closures with respect to all these topologies coincide. Any involutive subalgebra A of $\mathcal{L}(H)$, closed for these topologies and coinciding with its bicommutant $A^{\sim\sim}$, is called *von Neumann algebra* over H; then $\underset{\sim}{1} = \mathrm{id}_H$ belongs to A.

Every von Neumann algebra constitutes a C*-algebra. Conversely, any C*-algebra A may be considered to be a von Neumann algebra if and only if it admits a predual A_*; A_* is a Banach space with $A_* \subset A^*$([T] Chapter III Theorem 3.5 ; an elegant proof of this characterization of von Neumann algebras is performed by Kadison in [47]). If A is a von Neumann algebra over the Hilbert space H, one may describe A_* as the set of all $\omega = \sum_{n=1}^{\infty} \omega_{\xi_n, \eta_n}$, where (ξ_n), (η_n) are sequences in H for which $\sum_{n=1}^{\infty} \|\xi_n\|^2 < \infty$, $\sum_{n=1}^{\infty} \|\eta_n\|^2 < \infty$; A_* is the vector space consisting of the linear functionals on A that are ultraweakly continuous or, equivalently, ultrastrongly continuous. The elements of A_* are called *normal functionals* on A. This definition is consistent with the general definition of normality for the positive linear functionals on C*-algebras ([T] Chapter III, Corollary 3.11).

In particular, if H is a Hilbert space, $\mathcal{L}(H)$ is a von Neumann algebra. We have

$$(\mathcal{L}(H) \otimes_\gamma \mathcal{L}(H)_*)^* \simeq \mathcal{L}(\mathcal{L}(H), \mathcal{L}(H)) = \mathcal{L}(\mathcal{L}(H)).$$

The abelian von Neumann algebras are the algebras $L^\infty(\Omega)$ associated to locally compact spaces Ω equipped with positive measures, and considered to be multipliers over the corresponding Hilbert space $L^2(\Omega)$.

If A is a von Neumann algebra over the Hilbert space H, then A_* is also constituted by the elements $f \in A^*$ that are **-ultrastrongly continuous*, i.e., continuous with respect to the *-ultrastrong topology defined by the seminorms

$$x \mapsto (\omega(xx^*) + \omega(x^*x))^{\frac{1}{2}}$$

on A, where ω runs over the set of positive elements in $\mathcal{L}(H)_*$. If ω is a normal state on the von Neumann algebra A, we put

$$\|x\|_\omega^\# = \omega(\frac{xx^* + x^*x}{2})^{\frac{1}{2}}$$

for $x \in A$; $\|a\|_\omega^\# = 1$ for any unitary element a in A. The net (x_i) in A converges *-ultrastrongly to $x \in A$ if and only if $\lim_i \|x_i - x\|_\omega^\# = 0$ for every normal state ω on A.

Let A be a von Neumann algebra over the Hilbert space H and let $T \in \mathcal{L}(H)$. Then $T \in A$ if and only if $UTU^{-1} = T$ for every $U \in A^\sim_u$ ([D] Chapter I, §1, No.3, Proposition 3, Corollary); hence $T \in A^\sim$ if and only if $UTU^{-1} = T$ for every $U \in A^{\sim\sim\sim}_u = A_u$.

Let H be a Hilbert space and let A be a nonvoid subset of $\mathcal{L}(H)$ that is invariant by involution. Then $(A^{\sim\sim})^{\sim\sim} = A^{\sim\sim}$ is the smallest von Neumann algebra containing A; we say that $A^{\sim\sim}$ is the *von Neumann algebra generated by* A.

Let A be a C*-algebra. To any *-representation π of A on a Hilbert space H one associates the von Neumann algebra $\pi(A)^{\sim\sim}$ generated by $\pi(A)$. The representation π is said to be universal if, for every *-representation π_1

of A on the Hilbert space H_1, there exists a $*$-homomorphism ρ of $\pi(A)$ onto $\pi_1(A)$ that is continuous with respect to the ultraweak topologies and satisfies $\rho \circ \pi = \pi_1$. Then $\pi(A)^{\sim\sim}$ is the *enveloping von Neumann algebra* of A; $\pi(A)^{\sim\sim}$ may be identified with A** via an isometry that is a homeomorphism for the ultraweak topology of $\mathcal{L}(H)$ and the weak-*-topology of A** ([T] Chapter III, Theorem 2.4).

Let A be a von Neumann algebra over the Hilbert space H. The von Neumann subalgebra B of A is called *factor* of A in case $B \cap B^{\sim} = \underset{\sim}{C}\, id_H$.

Let A be a von Neumann algebra and let $T \in A$. There exists a minimal projection p in $A \cap A^{\sim}$ such that $Tp = T$ (and $pT = T$); p is called *central support* of T ([D] p.6).

Let H be a Hilbert space and let p be a projection of H onto X. If $T \in \mathcal{L}(H)$, we denote by T_p or T_X the restriction of pT to X; $T_X \in \mathcal{L}(X)$ and $T_X = (Tp)_X = (pT)_X = (pTp)_X$. If A is a subset of $\mathcal{L}(H)$ we denote by A_p or A_X the set $\{T_X : T \in A\}$.

Let A be a von Neumann algebra and let p be a projection in A. Then

$$(pAp)^{\sim} = pA^{\sim}p;$$

hence $(pAp)^{\sim\sim} = pA^{\sim\sim}p = pAp$; A_p is a von Neumann algebra. ([D] Chapter I, §2, Proposition 1; [T] Chapter II, Proposition 3.10).

If A is a von Neumann algebra, there exists a von Neumann algebra B and a projection p in $B^{\sim}$ such that $A \simeq B_p$ ([D] Chapter I, §4, Theorem 3, Corollary).

Two von Neumann algebras A_1, A_2 over the Hilbert spaces H_1, H_2 are said to be *spatially isomorphic* if there exists an isometry U of H_1 onto H_2 such that $UA_1 = A_2U$.

Consider two von Neumann algebras A and B. Any positive linear mapping f of A into B is called *normal* if for every increasing net (x_i) in A_+ converging to x, $(f(x_i))$ converges to f(x). Then the mapping f is continuous with respect to the ultraweak topologies. If, moreover, f transforms the unit of A into the unit of B, then f(A) is a von Neumann algebra ([D] Chapter I, §4, Theorem 2, Corollary 2). Given a Hilbert space H, any *-homomorphism of a von Neumann algebra A into $\mathcal{L}(H)$, that is uniformly continuous and also continuous with respect to the ultraweak topologies, constitutes a *normal representation.*

We call *Hilbert algebra* any involutive algebra A equipped with a scalar product defining a pre-Hilbert structure on A and satisfying the following conditions: a) For all $x,y \in A$, $(x|y) = (y^*|x^*)$; b) for all $x,y,z \in A$, $(xy|z) = (y|x^*z)$; c) for every $a \in A$, the left translation $\underset{A \to A}{x \mapsto ax}$ is continuous for the associated norm and, in the corresponding topology, $\{xy : x,y \in A\}^- = A$. Then, for every $a \in A$, the right translation $\underset{A \to A}{x \mapsto xa}$ is also continuous. The involution j on A admits an extended involution J on the Hilbert space completion H of A.

The von Neumann algebra A over the Hilbert space H is called *standard von Neumann algebra* if there exists a Hilbert algebra B such that $\bar{B} = H$ and A may be identified with the von Neumann algebra generated by the left translations on B.

If A is a standard von Neumann algebra and j is the involution in A, then the mapping $x \mapsto jx^*j$ is a linear antiisomorphism of A onto $A^{\sim}$; $jAj = A^{\sim}$ and $A^{\sim} = A^c$ ([D] Chapter I, §5, Proposition 2, Theorem 1).

Let A and B be von Neumann algebras. We may inject $A_* \otimes B_*$ canonically into $A^* \otimes B^*$. The closure $A_* \bar{\otimes} B_*$ of $A_* \otimes B_*$ in $(A \otimes_{min} B)^*$ is the predual of a von Neumann algebra that is denoted by $A \bar{\otimes} B$ and is called *tensor product of the von Neumann algebras* A and B ([T] p. 220-221). If A is a von Neumann algebra and H is a Hilbert space, then $(A \otimes (\underset{\sim}{C}\, id_H))^{\sim} = A^{\sim} \bar{\otimes} \mathcal{L}(H)$ ([T] Chapter IV, Proposition 1.6).

Let A_1, A_2, B_1, B_2 be von Neumann algebras and let ϕ_1 [resp. ϕ_2] be a normal morphism of A_1 into B_1 [resp. A_2 into B_2]. Then there exists a unique normal morphism ϕ of $A_1 \bar{\otimes} A_2$ into $B_1 \bar{\otimes} B_2$ such that $\phi(a_1 \otimes a_2) = \phi_1(a_1) \otimes \phi_2(a_2)$ whenever $a_1 \in A_1$, $a_2 \in A_2$ ([D] Chapter I, §4, Proposition 2; [T] Chapter IV, Theorem 5.2).

Let $(e_i)_{i \in I}$ be an orthonormal basis of the Hilbert space H. We put

$$TrT = \sum_{i \in I} (Te_i|e_i)$$

for $T \in \mathcal{L}(H)$; Tr is a normal, faithful, semifinite trace on $\mathcal{L}(H)_+$ and is independent of the choice of the orthonormal basis ([D] Chapter I, §6, Theorem 5).

If A is a von Neumann algebra over the Hilbert space H, one defines a *trace of* A *into a von Neumann subalgebra* B of A to be a mapping $\Phi : A_+ \to B$ such that a) $\Phi(x+y) = \Phi(x) + \Phi(y)$ whenever $x,y \in A_+$, b) $\Phi(\alpha x) = \alpha\Phi(x)$

whenever $\alpha \in \underset{\sim}{R}_+$ and $x \in A_+$, c) $\Phi(y^*xy) = y^*\Phi(x)y$ whenever $x \in A_+$, $y \in B_u$, d) $\Phi(x^*x) = \Phi(xx^*)$ whenever $x \in A$.

We summarize a collection of important results concerning crossed products. An excellent reference is [69].

One calls *covariant system* every triple (A,G,α), where A is a von Neumann algebra over a Hilbert space H, G is a locally compact group, and α is a continuous action of G on A, i.e., a homomorphism $s \mapsto \alpha_s$ of G into the group formed by the *-automorphisms of A, continuous with respect to the strong topology of A. For instance, if U is a continuous unitary representation of G on H, i.e., a continuous homomorphism of G into $\mathcal{L}(H)_u$, such that $U_s x \cdot U_s^* \in A$ whenever $x \in A$ and $s \in G$, one may choose $\alpha_s(x) = U_s x \cdot U_s^*$ $(s \in G, x \in A)$.

There exists a normal, faithful *-representation $x \mapsto \lambda_x$ of A on $L^2(G,H)$:

$$(\lambda_x f)(s) = \alpha_{s^{-1}}(x)f(s)$$

for $x \in A$, $f \in L^2(G,H)$, $s \in G$ ([23] Part I, Proposition 2.5). If $t \in G$, we consider also $\lambda'_t \in \mathcal{L}(L^2(G,H))$ defined by

$$(\lambda'_t f)(s) = f(t^{-1}s)$$

for $f \in L^2(G,H)$, $s \in G$; λ' is a continuous unitary representation of G on $L^2(G,H)$. For all $x \in A$ and $t \in G$,

$$\lambda'_t \lambda_x \lambda'^*_t = \lambda_{\alpha_t(x)}.$$

One defines the *crossed product* $A \otimes_\alpha G$ of A by G, with respect to α, to be the von Neumann algebra generated by $\{\lambda_x : x \in A\} \cup \{\lambda'_t : t \in G\}$. In the particular case where, for a continuous unitary representation U of G, $\alpha_s(x) = U_s x U_s^*$ $(s \in G, x \in A)$, the crossed product is spatially isomorphic to the von Neumann algebra generated by $\{x \otimes \underset{\sim}{1} : x \in A\} \cup \{U_t \otimes L_t : t \in G\}$, where $L_t\phi(s) = \phi(t^{-1}s)$ for $\phi \in L^2(G)$, $s \in G$, $t \in G$ ([69] Part I, Proposition 2.12).

We consider the continuous unitary representation R of G on $L^2(G)$ defined by

$$R_a\phi(s) = \phi(sa^{-1})\Delta(a^{-1})^{\frac{1}{2}}$$

for $\phi \in L^2(G)$, $a \in G$. If $a \in G$, we define $\rho_a \in \mathcal{L}(\mathcal{L}(L^2(G)))$ by

$$\rho_a(F) = R_{a^{-1}}FR_a,$$

$F \in \mathcal{L}(L^2(G))$. Considering the particular case of the action α, we also put

$$\alpha'_a = \alpha_a \otimes \rho_a,$$

$a \in G$. So we may define a continuous action α' of G on $A \bar{\otimes} \mathcal{L}(L^2(G))$; $A \otimes_\alpha G$ is the set of the elements in $A \bar{\otimes} \mathcal{L}(L^2(G))$ that are fixed for the action α' ([69] Part I, Theorem 3.11).

Let now G be abelian. The continuous unitary representations of G are the continuous homomorphisms of G into the torus $\{z \in \underset{\sim}{C}: |z| = 1\}$; they form the locally compact abelian dual group $\hat{G}$ of G ([40] I; p. 355). If $\chi \in \hat{G}$, we put

$$V_\chi f(a) = \overline{\chi(a)}f(a)$$

for $f \in L^2(G)$, $a \in G$, and

$$\hat{\alpha}_\chi(\xi) = (\underset{\sim}{1} \otimes V_\chi)\xi(\underset{\sim}{1} \otimes V_\chi)^*$$

for $\xi \in A \otimes_\alpha G$. So we define a dual action $\hat{\alpha}$ of $\hat{G}$ on $A \otimes_\alpha G$. The crossed product $(A \otimes_\alpha G) \otimes_{\hat{\alpha}} \hat{G}$ is isomorphic to $A \bar{\otimes} \mathcal{L}(L^2(G))$; $\{\lambda_x : x \in A\}$ is the set of elements in $A \otimes_\alpha G$ that are fixed under the action of $\hat{\alpha}$([69] Part I, Theorem 4.10, Proposition 4.12; [67] Chapter II, Theorem 2.2).

Let A be a von Neumann algebra and let f be a normal, faithful, positive linear functional on A. Then there exists a one-parameter group $\{\sigma^f_t : t \in \underset{\sim}{R}\}$ consisting of *-automorphisms of A, that are continuous for the strong topology, and satisfy the following property: If $x,y \in A$, there exists a bounded, continuous, complex-valued function F defined on $\mathcal{D} = \{\alpha+i\beta : \alpha \in \underset{\sim}{R}, \beta \in [0,1]\}$, analytic on $\mathcal{D}^0$, such that $F(t) = f(\sigma^f_t(x)y)$, $F(t+i) = f(y\sigma^f_t(x))$ whenever $t \in \underset{\sim}{R}$. We say that σ^f is the *modular action* associated to f; $\{\sigma^f_t : t \in \underset{\sim}{R}\}$ is the *modular group*. One considers the crossed product $A \otimes_{\sigma^f} \underset{\sim}{R}$.

If f and g are normal, faithful, positive linear functionals on A, then the von Neumann algebras $A \otimes_{\sigma^f} \underset{\sim}{R}$ and $A \otimes_{\sigma^g} \underset{\sim}{R}$ are isomorphic ([69] Part II, Theorem 2.1, Theorem 2.3; [67] Chapter I, Theorem 4.1).

If the von Neumann algebra A admits a normal trace θ, then there exists a unique central projection z in A such that the restriction of θ to Az is semifinite and $\theta(x) = \infty$ whenever $x \in A_+(\underset{\sim}{1}-z)$ with $z \neq 0$. The von Neumann algebra A is called *semifinite* if it admits a normal, faithful, semifinite trace on A_+. Every semifinite von Neumann algebra admits an isomorphism onto a standard von Neumann algebra ([D] Chapter I, §6, Theorem 2; Proposition 9, Corollary).

Let A be a semifinite factor over the Hilbert space H. The ideal m_θ associated to a normal, faithful, semifinite trace θ on A_+ is independent of the choice of θ. The elements of m_θ are called *tracial elements;* in the particular case $A = \mathcal{L}(H)$, they are called *tracial operators.* The Banach space formed by the tracial operators may be identified with the predual $\mathcal{L}(H)_*$ via the isometric linear mapping that associates to any tracial operator U the element ϕ_U of $\mathcal{L}(H)^*$ defined by

$$\langle T,\phi_U\rangle = Tr(TU) \; [= Tr(UT)],$$

$T \in \mathcal{L}(H)$ ([S] Proposition 1.15.3).

Let A be a von Neumann algebra. If θ is a normal, faithful trace on A_+, we define a norm $\|\cdot\|_2$ on A by the relation

$$\|x\|_2 = \theta(x^*x)^{\frac{1}{2}},$$

$x \in A$. For $m \in \underset{\sim}{\mathbb{N}}^*$, we consider the tracial state

$$t_m = \frac{1}{m} Tr$$

on $\mathfrak{m}_m(A)$. We put

$$\|y\|_2 = \|y\|^{\#}_{t_m} = t_m(y^*y)^{\frac{1}{2}},$$

$y \in \mathfrak{m}_m(A)$.

Two projections p and q in a von Neumann algebra A are said to be *equivalent* and one writes $p \sim q$ if there exists $a \in A$ such that $a^*a = p$ and $aa^* = q$; p is called *initial projection* of a and q is called *final projection* of a.

Let p be a projection in the von Neumann algebra A. One says that p is finite if, for every projection q in A such that $q \sim p$ and $q \leq p$, one has $p = q$. If p is not finite, it is called *infinite*. The projection p is termed *purely infinite* if it does not majorize a nonzero finite projection in A; it is termed *properly infinite* if, for every central projection q in A such that $qp \neq 0$, qp is infinite. The projection p is said to be *abelian* if the von Neumann algebra A_p is abelian. Every abelian projection is finite.

The *von Neumann algebra* A is called *finite, infinite, purely infinite, properly infinite* in case the corresponding projection id_A is. Every purely infinite von Neumann algebra is properly infinite. The von Neumann algebra is said to be of *type* I if every nonzero central projection majorizes a nonzero abelian projection; a factor of type I is of the form $\mathcal{L}(H)$ for a Hilbert space H. The von Neumann algebra is said to be of type II if it does not admit a nonzero abelian projection and every nonzero central projection majorizes a nonzero finite projection. The von Neumann algebra of type II is called of *type* II_1 if it is finite, of *type* II_∞ if it does not admit a nonzero central finite projection. Finally a von Neumann algebra is said to be of *type* III if it is purely infinite, i.e., it does not admit a nonzero finite projection.

Every von Neumann algebra admits a unique decomposition into a direct sum of von Neumann algebras of types I, II_1, II_∞, III, as well as into a direct sum of a finite von Neumann algebra and a properly infinite von Neumann algebra ([T] Chapter V, Theorem 1.19). If the von Neumann algebra is not of type III, it is semifinite ([T] Chapter V, Theorem 2.15). If A is a properly infinite von Neumann algebra and H is a separable Hilbert space, then $A \bar{\otimes} \mathcal{L}(H) \simeq A$ ([69] Appendix C).

Every abelian von Neumann algebra is finite. For every finite von Neumann algebra A, I(A) coincides with A_u. Every von Neumann algebra A admits a maximal central projection p for which A_p is a finite von Neumann algebra ([D] p. 98; [T] Chapter V, Theorem 1.19). The von Neumann algebra A is finite if and only if it admits a linear (ultraweakly continuous) operator T of A onto the center Z of A satisfying the following properties:
a) $T(xx^*) = T(x^*x) \geq 0$ whenever $x \in A$; b) $T(ax) = aT(x)$ whenever $x \in A$, $a \in Z$,

c) $\tilde{T}(1) = 1$; d) $\tilde{T}(x^*x) > 0$ for every $x \in A$, $x \neq 0$ ([T] Chapter V, Theorem 2.6). We call $\tilde{T}$ a *central trace*.

If A is a von Neumann algebra and f is a normal, faithful, positive linear functional on A, then the crossed product $A \otimes_{\sigma^f} \tilde{R}$ constitutes a semifinite von Neumann algebra ([69] Part II, Section 3; [67] Chapter II, Corollary 1.11). Let A be a von Neumann algebra of type III. Then A is the crossed product of a von Neumann algebra of type II_∞ by $\tilde{R}$; the crossed product is unique if one considers modular actions ([69] Part II, Theorem 4.8; [67] Chapter III, Theorem 1.1).

A. Amenability properties on Banach algebras and C*-algebras related to von Neumann algebras.

If G is a locally compact group, one considers the left regular representation $x \mapsto L_x$ defined by $L_x f = {}_{x^{-1}}f$ for $f \in L^2(G)$. In $\mathcal{L}(L^2(G))$, the C*-algebra $G \to \mathcal{L}(L^2(G))$

generated by $\{L_x : x \in G\}$ is denoted by $C_L^*(G)$ and the von Neumann algebra, called algebra of pseudomeasures, generated by $\{L_x : x \in G\}$ is denoted by PM(G). As not every discrete group G is amenable ([P] 14), the following proposition establishes the existence of nonamenable C*-algebras.

PROPOSITION 3.1: Let G be a discrete group.

1) If A is an amenable C*-algebra such that $C_L^*(G) \subset A \subset PM(G)$, then G is an amenable group.

2) If G is an amenable group, then $C_L^*(G)$ is an amenable C*-algebra.

PROOF: As G is discrete, we denote $L^2(G)$ by $\ell^2(G)$.

1') We may consider $\mathcal{L}(\ell^2(G))$ to be a Banach A-module admitting A as a Banach A-submodule. Let $F \in A^*$ be defined by

$$\langle a, F\rangle = (a\delta_e | \delta_e),$$

$a \in A$. For all $a \in A$ and $v \in A_u$, we have

$$F(vav^*) = (vav^*\delta_e | \delta_e) = (av^*\delta_e | v^*\delta_e) = (a\delta_e | \delta_e) = F(a).$$

Moreover, $F(id_{\ell^2(G)}) = 1$. By Proposition 2.4 there exists $F' \in \mathcal{L}(\ell^2(G))$ extending F such that $F'(vav^*) = F'(a)$ whenever $a \in \mathcal{L}(\ell^2(G))$ and $v \in A_u$. We consider the adjoint linear functional $F_1' : a \mapsto \overline{F'(a^*)}$ in $\mathcal{L}(\ell^2(G))^*$ and also the hermitian linear functional $F'' = F' + F_1' \in \mathcal{L}(\ell^2(G))^*$. There exists a unique pair (F_1'', F_2'') of positive linear functionals in $\mathcal{L}(\ell^2(G))^*$ such that $F'' = F_1'' - F_2''$, $\|F''\| = \|F_1''\| + \|F_2''\|$ and also $F_i''(vav^*) = F_i''(a)$ whenever $a \in \mathcal{L}(\ell^2(G))$, $v \in A_u$; $i = 1,2$. As F' extends F, F_1'' and F_2'' are not both the null functional. Hence there exists a positive linear functional Φ in $\mathcal{L}(\ell^2(G))^*$ such that $\Phi(id_{\ell^2(G)}) = 1$ and $\Phi(vav^*) = \Phi(a)$ whenever $a \in \mathcal{L}(\ell^2(G))$ and $v \in A_u$; in particular, $\Phi(L_{x^{-1}} a L_{x^{-1}}^*) = \Phi(a)$, whenever $a \in \mathcal{L}(\ell^2(G))$ and $x \in G$.

If $f \in \ell^\infty(G)$, we define $M_f \in \mathcal{L}(\ell^2(G))$ by putting

$$M_f \phi = f\phi$$

for $\phi \in \ell^2(G)$. If $f \in \ell_+^\infty(G)$, M_f is a positive element in $\mathcal{L}(\ell^2(G))$. For the unit function 1_G on G, we have $M_{1_G} = id_{\ell^2(G)}$. For all $f \in \ell^\infty(G)$, $\phi \in \ell^2(G)$ and $x,z \in G$,

$$L_{x^{-1}} M_f L_{x^{-1}}^* \phi(z) = L_{x^{-1}} M_f L_x \phi(z) = L_{x^{-1}} f(z)\phi(x^{-1}z)$$

$$= f(xz)\phi(z) = ({}_x f\phi)(z);$$

hence

$$L_{x^{-1}} M_f L_{x^{-1}}^* = M_{{}_x f}.$$

We finally define $M \in \ell^\infty(G)^*$ by putting

$$M(f) = \Phi(M_f),$$

$f \in \ell^\infty(G)$. If $f \in \ell_+^\infty(G)$, then $M(f) \geq 0$. Moreover, $M(1_G) = \Phi(id_{\ell^2(G)}) = 1$. For all $f \in \ell^\infty(G)$ and $x \in G$,

$$M({}_x f) = \Phi(M_{{}_x f}) = \Phi(L_{x^{-1}} M_f L_{x-1}^*) = \Phi(M_f) = M(f).$$

Thus M is a left invariant mean on $\ell^\infty(G)$ and G is amenable.

2) Let X be a Banach $C_L^*(G)$-module and let Y be a Banach submodule of X. Consider $F \in Y^*$ such that $F(v\xi v^*) = F(\xi)$ whenever $\xi \in Y$ and $v \in (C_L^*(G))_u$. Let F' be an extension of F to an element in X* and put

$$F_x(\xi) = F'(L_x\xi L_x{}^*)$$

for $x \in G$, $\xi \in X$. If $\xi \in X$, we may consider $F'_\xi \in \ell^\infty(G)$ defined by

$$F'_\xi(x) = F_x(\xi),$$

$x \in G$. As G is amenable, there exists a right invariant mean M on $\ell^\infty(G)$, i.e., for $\ell^\infty(G)$ and $x \in G$, $M(f) = M(f_x)$. Let also

$$\Phi : \xi \mapsto M(F'_\xi);$$
$$X \to \underset{\sim}{C}$$

then $\Phi \in X^*$. For every $\eta \in Y$ and every $x \in G$,

$$F_x(\eta) = F'(L_x\eta L_x{}^*) = F(L_x\eta L_x{}^*) = F(\eta).$$

Hence $F'_\eta(x) = F(\eta)$ and $\Phi(\eta) = F(\eta)$, i.e., Φ extends F.

If $\xi \in X$ and $x \in G$, $a \in G$, we have

$$F'_{L_a\xi L_a{}^*}(x) = F_x(L_a\xi L_a{}^*) = F'(L_xL_a\xi L_a{}^*L_x{}^*)$$
$$= F'(L_{xa}\xi L_{xa}{}^*).$$

As M is right invariant,

$$\Phi(L_a\xi L_a{}^*) = M(F'_{L_a\xi L_a{}^*}) = M(F'_\xi) = \Phi(\xi);$$

hence also

$$\Phi(L_a\xi) = \Phi(L_a\xi L_aL_a{}^*) = \Phi(\xi L_a).$$

Thus, for all $\xi \in X$ and $v \in (C_L^*(G))_u$,

$$\Phi(v\xi) = \Phi(\xi v)$$

and

$$\Phi(v\xi v^*) = \Phi(\xi v^* v) = \Phi(\xi).$$

The property now follows from Proposition 2.4. □

We add a result concerning strongly amenable C*-algebras.

PROPOSITION 3.2: If T is a representation of a strongly amenable C*-algebra A on a Hilbert space H, there exist a positive-definite operator S in $\mathcal{L}(H)$ and a *-representation T' of A on H such that $T'_a = S^{-1}T_aS$ whenever $a \in A$.

PROOF: It suffices to consider the case of a unital strongly amenable C*-algebra.

We equip $\mathcal{L}(H)$ with the dual Banach A-module structure defined by

$$aV = T_aV, \quad Va = V(T_{a^*})^*,$$

$a \in A$, $V \in \mathcal{L}(H)$. Consider $D \in \mathcal{L}(A, \mathcal{L}(H))$ defined by

$$D(a) = T_a - (T_{a^*})^*,$$

$a \in A$. For all $a, b \in A$,

$$aD(b) + D(a)b = T_a(T_b-(T_{b^*})^*) + (T_a-(T_{a^*})^*)(T_{b^*})^*$$

$$= T_{ab} - (T_{a^*})^*(T_{b^*})^* = T_{ab} - (T_{b^*a^*})^*$$

$$= T_{ab} - (T_{(ab)^*})^* = D(ab);$$

D is a continuous derivation. As A is strongly amenable, there exists an element V in the closed convex hull of $\{D(x)x^*: x \in A_u\}$ such that $D = -d_1V$. For every $a \in A$, we have.

$$T_a - (T_{a*})^* = D(a) = -aV + Va = -T_aV + V(T_{a*})^*;$$

hence

$$T_a(\mathrm{id}_H + V) = (\mathrm{id}_H + V)(T_{a*})^*. \qquad (1)$$

Let $U = \mathrm{id}_H + V$. For every $v \in A_u$, we have

$$D(v)v^* = (T_v - (T_{v*})^*)\, T_v^* = T_vT_v^* - \mathrm{id}_H$$

and U belongs to the closed convex hull of $\{T_vT_v^* : v \in A_u\}$.

If $v \in A_u$ and $\xi \in H$, $(T_vT_v^*\xi|\xi) \geqq 0$. Hence U is a positive-definite operator. Let S be the square root of U.

For all $\xi \in H$ and $v \in A_u$,

$$\|\xi\| = \|(T_vT_{v*})^*\xi\| = \|(T_{v*})^*T_v^*\xi\| \leqq \|T\| \; \|T_v^*\xi\| . \qquad (2)$$

Let $\xi \in H$ such that $U\xi = 0$. For every $\varepsilon > 0$, we may determine $\alpha_1,\ldots,\alpha_n \in \underset{\sim}{R}^*_+$ and $v_1,\ldots,v_n \in A_u$ such that $\sum_{i=1}^{n} \alpha_i = 1$ and $\|U - \sum_{i=1}^{n} \alpha_i T_{v_i}T_{v_i}^*\| < \varepsilon$; hence

$$\sum_{i=1}^{n} \alpha_i \|T_{v_i}^*\xi\|^2 = \sum_{i=1}^{n} \alpha_i(T_{v_i}T_{v_i}^*\xi|\xi)$$

$$\leqq \varepsilon\|\xi\|^2 + (U\xi|\xi) = \varepsilon\|\xi\|^2.$$

Thus, for every $\varepsilon > 0$, there exists $v \in A_u$ such that $\|T_v^*\xi\| \leqq \varepsilon^{1/2}\|\xi\|$. From (2) we deduce that $\xi = 0$. Therefore U^{-1} exists; then S^{-1} exists too.

The relation (1) expresses that, for every $a \in A$,

$$T_aS^2 = S^2(T_{a*})^*,$$

$$S^{-1}T_aS = S(T_{a*})^*S^{-1}.$$

For $a \in A$, let $T'_a = S^{-1}T_aS$; then

$$(T_a')^* = (S^{-1}T_aS)^* = (S(T_{a^*})^*S^{-1})^* = S^{-1}T_{a^*}S = T_{a^*}'. \qquad \square$$

PROPOSITION 3.3: Let H be a Hilbert space and let A be a *-subalgebra of $\mathcal{L}(H)$ that is strongly amenable. Consider $T \in \mathcal{L}(H)$ such that $aT(H) \subset T(H)$ whenever $a \in A$. Then there exists $T' \in A^{\sim}$ such that $T'(H) = T(H)$.

PROOF: For every $\xi \in H$,

$$\|T\xi\|^2 = (T^*T\xi|\xi) = (|T|^2\xi|\xi) = (|T|\xi||T|\xi) = \||T|\xi\|^2.$$

Hence $Ker\ T = Ker\ |T|$; T and $|T| = (T^*T)^{\frac{1}{2}}$ admit the same image. We may then suppose T to be positive-definite.

If $a \in A$ and $\xi \in X$, by hypothesis, there exists $\xi_0 \in H$ such that $aT(\xi) = T(\xi_0)$. There exists a unique vector $\eta = \eta_{a,\xi} \in (Ker\ T)^\perp$ such that $T(\xi_0) = T(\eta)$, i.e.

$$aT(\xi) = T(\eta_{a,\xi}).$$

We may then define a representation U of A on H by putting

$$U_a(\xi) = \eta_{a,\xi}$$

for $a \in A$, $\xi \in H$. If $a \in A$, $U_a \in \mathcal{L}(H)$; moreover, $U_a(H) \subset (Ker\ T)^\perp = Im\ T$ and

$$aT = TU_a. \qquad (3)$$

By Proposition 3.2 one may determine a positive-definite operator S in $\mathcal{L}(H)$ such that $W: a \mapsto S^{-1}U_aS$ is a *-representation of A. Consider the polar decomposition $T''T'$ of ST, where $T' = |ST|$ and T" is a partially isometric operator. We have

$$T'T''^* = T'^*T''^* = T^*S = TS;$$

hence $T = T'T''^*S^{-1}$ and $T(H) \subset T'(H)$. On the other hand, $TST'' = T'T''^*T''$ and $T'(H) \subset T(H)$.

For every $a \in A$, by (3), $aT = TSW_aS^{-1}$; hence $aTS = TSW_a$ and

$$aT'T''* = aTS = TSW_a = T'T''*W_a,$$

$$aT' = T'T''*W_aT''.$$

Then we have also

$$a*T' = T'T''*W_{a*}T'' = T'T''*W_a*T''$$

and

$$T'a = T''*W_aT''T'.$$

Thus

$$T'^2a = T'T''*W_aT''T' = aT'^2;$$

$T'^2 \in A^\sim$. As the von Neumann algebra $A^\sim$ is a C*-subalgebra of $\mathcal{L}(H)$ and T' is positive-definite, we conclude that $T' = (T'^2)^{1/2} \in A^\sim$. □

We now consider a unilateral amenability-type property for certain Banach algebras.

We call *Lau algebra* any pair (A,B) consisting of a Banach algebra A and a von Neumann algebra B for which $A = B_*$ (and A* = B) and the unit u_B of B is a multiplicative linear functional on A. The Lau algebra is also denoted simply by A although, in general, B is not necessarily unique. If G is a locally compact group, the group algebra $L^1(G)$, the Fourier algebra A(G), the Fourier-Stieltjes algebra B(G) ([P] 10.B, 19.A) are Lau algebras ([54]).

DEFINITION 3.4: The Lau algebra (A,B) is called left amenable if, for every Banach A-module X such that

$$a \cdot \xi = \langle a, u_B \rangle \xi$$

whenever $a \in A$ and $\xi \in X$, one has $H_1(A, X^*) = \{0\}$.

Let (A,B) be a Lau algebra and let $P_1(A) = \mathfrak{m}(A^*) \cap A$. If $a,b \in A$, the element ab of A may be considered to be a continuous linear functional on A^*. If $f \in A^*$, $f \geqq 0$, then

$$\langle f,ab\rangle = f(ab) \geqq 0$$

and

$$\langle u_B,ab\rangle = u_B(ab) = u_B(a)u_B(b) = 1;$$

hence $P_1(A)$ is a semigroup.

We call *right stable mean* on A^* any mean M on the Banach A-module A^* such that

$$\langle f\phi,M\rangle = \langle f,M\rangle$$

whenever $f \in A^*$ and $\phi \in P_1(A)$.

<u>PROPOSITION 3.5</u>: The Lau algebra A is left amenable if and only if A^* admits a right stable mean.

a) Suppose the existence of a right stable mean M on A^*. Consider a Banach A-module X such that

$$a\cdot\xi = \langle a,u_B\rangle\xi$$

for all $a \in A$ and $\xi \in X$. Let D be a continuous derivation of A into X^*; then ${}^tD \in \mathcal{L}(X^{**},A^*)$. We put $D' = {}^tD|_X$; then ${}^tD' \in \mathcal{L}(A^{**},X^*)$ and $F = {}^tD'(M) \in X^*$.

Let $a \in A$ and $\phi \in P_1(A)$, $\xi \in X$. We have

$$\langle a,D'(\phi\cdot\xi)\rangle = \langle\phi\cdot\xi,D(a)\rangle = \langle\phi,u_B\rangle\langle\xi,D(a)\rangle$$

$$= \langle\xi,D(a)\rangle = \langle a,D'(\xi)\rangle;$$

hence

$$D'(\phi\cdot\xi) = D'(\xi)$$

and

$$\langle \xi, F\cdot\phi\rangle = \langle \phi\cdot\xi, F\rangle = \langle D'(\phi\cdot\xi), M\rangle$$

$$= \langle D'(\xi), M\rangle = \langle \xi, F\rangle,$$

i.e.,

$$F.\phi = F. \tag{4}$$

Moreover,

$$\langle a, D'(\xi\cdot\phi)\rangle = \langle \xi\cdot\phi, D(a)\rangle = \langle \xi, \phi\cdot D(a)\rangle$$

$$= \langle \xi, D(\phi a) - D(\phi)\cdot a\rangle$$

$$= \langle \phi a, D'(\xi)\rangle - \langle a\cdot\xi, D(\phi)\rangle$$

$$= \langle a, D'(\xi)\phi\rangle - \langle a, u_B\rangle \langle \xi, D(\phi)\rangle;$$

hence

$$D'(\xi\cdot\phi) = D'(\xi)\phi - \langle \xi, D(\phi)\rangle u_B.$$

As M is a right stable mean, we obtain now

$$\langle \xi, \phi\cdot F\rangle = \langle \xi\cdot\phi, F\rangle = \langle D'(\xi\cdot\phi), M\rangle$$

$$= \langle D'(\xi)\phi, M\rangle - \langle \xi, D(\phi)\rangle \langle u_B, M\rangle$$

$$= \langle D'(\xi), M\rangle - \langle \xi, D(\phi)\rangle$$

$$= \langle \xi, F - D(\phi)\rangle,$$

i.e.,

$$D(\phi) = F-\phi\cdot F. \tag{5}$$

By (4) and (5), for every $\phi \in P_1(A)$, we have

$$D(\phi) = \phi\cdot(-F) - (-F)\cdot\phi.$$

As $P_1(A)$ generates A^*, we conclude that $D = d_1(-F)$.

2) We suppose A to be left amenable.

If $a \in A$ and $f \in A^*$, we put

$$a\cdot f = \langle a,u_B\rangle f. \qquad (6)$$

As u_B is a multiplicative continuous linear functional, we may consider A^* to be a Banach A-module for the left action defined by (6) and the right action defined canonically.

Let $x \in A$ and $a \in A$. By (6) we have

$$\langle x,a\cdot u_B\rangle = \langle a,u_B\rangle \langle x,u_B\rangle;$$

as u_B is multiplicative, we have also

$$\langle x,u_B a\rangle = \langle ax,u_B\rangle = \langle a,u_B\rangle \langle x,u_B\rangle.$$

Hence

$$a\cdot u_B = u_B a = \langle a,u_B\rangle u_B;$$

$\underset{\sim}{C}u_B$ is a Banach A-submodule of A^*. We consider the Banach A-module $X = A^*/\underset{\sim}{C}u_B$ and let π be the canonical homomorphism of A^* onto X; ${}^t\pi$ is a morphism of the Banach A-module X^* into the Banach A-module A^{**} associated to A, and ${}^t\pi$ is injective.

Choose $F_o \in A^{**}$ such that $\langle u_B,F_o\rangle = 1$. If $a \in A$, $d_1F_o(a) \in A^{**}$ and

$$\langle u_B,d_1F_o(a)\rangle = \langle u_B,a \odot F_o - F_o\cdot a\rangle$$

$$= \langle u_B a - a\cdot u_B,F_o\rangle = 0.$$

Hence $d_1F_o(a)$ may be considered to be an element of ${}^t\pi(X^*)$. As ${}^t\pi$ is injective, one may define

$$D : A \to X^*$$

by putting

$$^t\pi \circ D(a) = d_1 F_o(a)$$

for $a \in A$; D is a continuous derivation. As A is left amenable, there exists $\Phi \in X^*$ such that, for every $a \in A$,

$$D(a) = a \cdot \Phi - \Phi a;$$

thus

$$a \odot ({}^t\pi(\Phi)) - ({}^t\pi(\Phi)) \cdot a = {}^t\pi(a \cdot \Phi - \Phi a) = {}^t\pi(D(a))$$

$$= d_1 F_o(a) = a \odot F_o - F_o \cdot a$$

and

$$a \odot ({}^t\pi(\Phi) - F_o) = ({}^t\pi(\Phi) - F_o) \cdot a.$$

Let $F = F_o - {}^t\pi(\Phi) \in A^{**}$. Then

$$\langle u_B, F\rangle = \langle u_B, F_o\rangle - \langle u_B, {}^t\pi(\Phi)\rangle$$

$$= \langle u_B, F_o\rangle - \langle \pi(u_B), \Phi\rangle = 1$$

and, for every $a \in A$, $a \odot F = F \cdot a$. Thus, for all $a \in A$ and $f \in A^*$,

$$\langle f, a \odot F\rangle = \langle f, F \cdot a\rangle = \langle a \cdot f, F\rangle = \langle a, u_B\rangle \langle f, F\rangle. \tag{7}$$

In particular, let $a \in P_1(A)$. For every $f \in A^*$, by (7), we have

$$\langle f, a \odot F\rangle = \langle f, F\rangle,$$

i.e.,

$$a \odot F = F.$$

We have also $a \odot F^* = F^*$ and we may thus suppose that F is hermitian. Then, $a \odot F_+$, $a \odot F_-$ are positive linear functionals; $a \odot F = a \odot F_+ - a \odot F_-$ and

$$\|a \odot F_+\| + \|a \odot F_-\| = \langle u_B, a \odot F_+\rangle + \langle u_B, a \odot F_-\rangle$$

$$= \langle u_B a, F_+\rangle + \langle u_B a, F_-\rangle$$

$$= \langle a, F_+\rangle + \langle a, F_-\rangle$$

$$= \|F_+\| + \|F_-\| = \|F\| .$$

So necessarily $a \odot F_+ = F_+$, $a \odot F_- = F_-$. As $F \neq 0$, the linear functionals F_+, F_- cannot both be null. Assume, for instance, that $F_+ \neq 0$; then $M = \frac{1}{\langle u_B, F_+\rangle} F_+$ is a mean on A*. For all $f \in A^*$ and $a \in P_1(A)$,

$$\langle fa, F_+\rangle = \langle f, a \odot F_+\rangle = \langle f, F_+\rangle.$$

Therefore M is right stable. □

The next proposition characterizes left amenability of a Lau algebra in terms of the Day-type and Glicksberg-Reiter-type properties describing amenability on the locally compact groups ([P] 6.A; 6.D).

PROPOSITION 3.6: Let A be a Lau algebra. The following conditions are equivalent:

(i) A is left amenable.

(ii) There exists a net $(\phi_i)_{i \in I}$ in $P_1(A)$ such that

$$\lim_i \|a\phi_i\| = |\langle a, u_B\rangle| \text{ whenever } a \in A.$$

(iii) For every $a \in \mathit{Ker}\ u_B$ and every $\varepsilon > 0$, there exists $\phi \in P_1(A)$ such that $\|a\phi\| < \varepsilon$.

PROOF: (i) => (ii).

With respect to the topology $\sigma(A^{**}, A^*)$, $P_1(A)$ is dense in $\mathfrak{m}(A^*)$; as $\mathfrak{m}(A^*)$ is convex, $P_1(A)$ is also dense with respect to the norm topology.

By Proposition 3.5 there exists a net $(\phi_i)_{i\in I}$ in $P_1(A)$ such that

$$\lim_i \|\phi\phi_i - \phi_i\| = 0$$

whenever $\phi \in P_1(A)$.

Let $\alpha_1,\ldots,\alpha_n \in \underset{\sim}{C}^*$ and $\psi_1,\ldots,\psi_n \in P_1(A)$; for $a = \sum_{p=1}^{n} \alpha_p \psi_p \in A$, we have $|\langle a,u_B\rangle| = |\sum_{p=1}^{n} \alpha_p|$. If $\varepsilon > 0$, we may determine $i_0 \in I$ such that, for $i \in I$, $i_0 \prec i$, we have

$$\|\psi_p \phi_i - \phi_i\| < \varepsilon /(n\, |\alpha_p|\,)$$

whenever $p = 1,\ldots,n$. Then

$$\|a\phi_i\| \leqq \|\sum_{p=1}^{n} \alpha_p \psi_p \phi_i - \sum_{p=1}^{n} \alpha_p \phi_i\| + \|\sum_{p=1}^{n} \alpha_p \phi_i\|$$

$$\leqq \sum_{p=1}^{n} |\alpha_p|\ \|\psi_p \phi_i - \phi_i\| + |\sum_{p=1}^{n} \alpha_p|$$

$$< \varepsilon + |\langle a,u_B\rangle|.$$

On the other hand,

$$|\langle a,u_B\rangle| = |\langle a\phi_i,u_B\rangle| \leqq \|a\phi_i\|.$$

Hence

$$0 \leqq \|a\phi_i\| - |\langle a\phi_i,u_B\rangle| < \varepsilon.$$

As $\varepsilon > 0$ may be chosen arbitrarily and $P_1(A)$ generates the vector space A, we proved (ii).

(ii) => (iii)

Trivial

(iii) => (i).

Let $\theta_0 \in P_1(A)$ be fixed.

Consider $\varepsilon \in \underset{\sim}{R}_+^*$ and a finite subset $F = \{\phi_1,\ldots,\phi_n\}$ of $P_1(A)$. We put $\psi_1 = \phi_1\theta_0 - \theta_0$. As u_B is a multiplicative linear functional, $\psi_1 \in \mathit{Ker}\ u_B$. Thus, by hypothesis, there exists $\theta_1 \in P_1(A)$ such that $\|\psi_1\theta_1\| < \varepsilon$. Now $\psi_2 = \phi_2\theta_0\theta_1 - \theta_0\theta_1 \in \mathit{Ker}\ u_B$ and there exists $\theta_2 \in P_1(A)$ such that $\|\psi_2\theta_2\| < \varepsilon$.

Continuing this procedure, for $i = 1,\ldots,n$, we determine $\theta_i \in P_1(A)$ such that $\|\psi_i\theta_i\| < \varepsilon$ with

$$\psi_i = \phi_i\theta_0\ \theta_1 \ldots \theta_{i-1} - \theta_0\ \theta_1 \ldots \theta_{i-1}.$$

We put $\eta = \eta_{F,\varepsilon} = \theta_0\theta_1 \ldots \theta_n$. Then

$$\|\phi_i\eta - \eta\| < \varepsilon$$

for every $i = 1,\ldots,n$.

We consider the net $(\eta_{F,\varepsilon})$, where F runs over the set of all finite subsets in $P_1(A)$ and $\varepsilon \in \underset{\sim}{R}_+^*$; for two such subsets F_1, F_2 and $\varepsilon_1,\varepsilon_2 \in \underset{\sim}{R}_+^*$, $(F_1,\varepsilon_1) \prec (F_2,\varepsilon_2)$ if and only if $F_1 \subset F_2$ and $\varepsilon_2 < \varepsilon_1$. Every limit point of $(\eta_{F,\varepsilon})$ with respect to the $\sigma(A^{**},A^*)$-topology is a right stable mean on A^*. Left amenability of A follows from Proposition 3.5. □

We indicate a transposition to Lau algebras of an amenability characterization for locally compact groups ([P] Proposition 6.17).

PROPOSITION 3.7: Let A be a Lau algebra. The following properties are equivalent:

(i) A is left amenable.

(ii) If I_1 and I_2 are right ideals in the semigroup $P_1(A)$, then $\inf\ \{\|a_1-a_2\| : a_1 \in I_1,\ a_2 \in I_2\} = 0$.

(iii) The set N(A) of all $f \in A^*$ for which $\inf\ \{\ \|f\phi\|\ : \phi \in P_1(A)\} = 0$ is closed under addition.

PROOF: (i) => (ii).

In the convex set $\mathfrak{M}(A^*)$, $P_1(A)$ is dense with respect to the $\sigma(A^{**},A^*)$-

topology, hence also with respect to the norm-topology. There exists a net (ϕ_i) in $P_1(A)$ such that $\lim_i \|\phi\phi_i - \phi_i\| = 0$ whenever $\phi \in P_1(A)$. Then, for $a_1 \in I_1$ and $a_2 \in I_2$,

$$\lim_i \|a_1\phi_i - a_2\phi_i\| \leqq \lim_i \|a_1\phi_i - \phi_i\| + \lim_i \|a_2\phi_i - \phi_i\| = 0.$$

(ii) => (iii).

Let $f_1, f_2 \in N(A)$ and $\varepsilon > 0$. There exist $a_1, a_2 \in P_1(A)$ such that $\|f_1a_1\| < \varepsilon$ and $\|f_2a_2\| < \varepsilon$. By hypothesis, one may determine $\phi_1, \phi_2 \in P_1(A)$ such that $\|a_1\phi_1 - a_2\phi_2\| < \varepsilon$. Then

$$\|(f_1+f_2)a_1\phi_1\| \leqq \|(f_1a_1)\phi_1\| + \|f_2(a_1\phi_1 - a_2\phi_2)\| + \|(f_2a_2)\phi_2\|$$

$$\leqq \varepsilon\|\phi_1\| + \varepsilon\|f_2\| + \varepsilon\|\phi_2\| = \varepsilon(2 + \|f_2\|).$$

As $\varepsilon > 0$ may be chosen arbitrarily,

$$\inf\{\|(f_1 + f_2)\phi\| : \phi \in P_1(A)\} = 0.$$

(iii) => (i).

The set $N(A)$ constitutes a vector subspace of A^*.
Let $\phi \in P_1(A)$ and $f \in A^*$. For $n \in \underset{\sim}{N}^*$, let $\phi_n = \frac{1}{n} \sum_{i=1}^{n} \phi^i \in P_1(A)$; then

$$\|(f\phi - f)\phi_n\| = \frac{1}{n}\|f(\phi^{n+1} - \phi)\| \leqq \frac{2}{n}\|f\|$$

and

$$\lim_{n\to\infty} \|(f\phi - \phi)\phi_n\| = 0.$$

Hence $f\phi - \phi \in N(A)$.

Consider the real vector subspace E of the hermitian linear functionals in A^* and let C be the convex set of all $f \in E$ such that $\inf\{\langle\phi, f\rangle : \phi \in P_1(A)\} > 0$; C is open in E, $u_B \in C$ and $C \cap N(A) = \emptyset$. By the Hahn-Banach theorem, there exists a real continuous linear functional Φ on E such that $\Phi(u_B) = 1$ and $\Phi(f) = 0$ for every $f \in E \cap N(A)$. Thus

$\Phi(f\phi - f) = 0$ whenever $f \in E$ and $\phi \in P_1(A)$. We put

$$\Psi(f) = \Phi(f_1) + i\Phi(f_2)$$

for $f = f_1 + if_2$, where $f_1, f_2 \in E$; $\Psi \in A^{**}$, $\Psi(u_B) = 1$, $\Psi \neq 0$. For all $f \in A^*$, $\phi \in P_1(A)$, we have $\Psi(f\phi) = \Psi(f)$ and also $\Psi^*(f\phi) = \Psi^*(f)$ for the conjugate functional Ψ^* of Ψ. Therefore we may suppose Ψ to be hermitian. One at least of the functionals Ψ_+, Ψ_- is not the null functional. If, for instance, $\Psi_+ \neq 0$, then $\frac{1}{\langle u_B, \Psi_+ \rangle} \Psi_+$ is a right stable mean on A^*; by Proposition 3.5, A is left amenable. □

In parts B, C we suppose separability for the Hilbert spaces on which von Neumann algebras are considered.

B. Semidiscrete von Neumann algebras

An important problem is the comparison of the norms on tensor products of unital C*-algebras. We study a certain number of equivalent properties characterizing the von Neumann algebras A for which the mapping ϖ, defined on $A \otimes A^{\sim}$, may be extended continuously on $A \otimes_{\min} A^{\sim}$.

DEFINITION 3.8: The von Neumann algebra A over the Hilbert space H is called semidiscrete, if id_A is a limit, with respect to the weak operator topology, of normal morphisms of A into itself having finite rank.

If H is a (separable) Hilbert space, $\mathcal{L}(H)$ is a semidiscrete von Neumann algebra. As a matter of fact, let $(p_i)_{i \in I}$ be an increasing net determined by all the finite-rank projections in $\mathcal{L}(H)$ and let θ be an arbitrary normal state on $\mathcal{L}(H)$. If $i \in I$, we put

$$\Phi_i(a) = p_i a p_i + \theta(a)(id_A - p_i),$$

$a \in \mathcal{L}(H)$; Φ_i is a finite-rank normal morphism of A into itself. The net $(\Phi_i)_{i \in I}$ converges simply to $id_{\mathcal{L}(H)}$.

A semidiscrete von Neumann algebra is not purely infinite, i.e., not of type III. Hence it is semifinite.

PROPOSITION 3.9: Let A be a semidiscrete von Neumann algebra and let p be a projection in A. Then A_p is also a semidiscrete von Neumann algebra.

PROOF: By hypothesis there exists a net $(\Phi_i)_{i\in I}$ of normal finite-rank morphisms of A into itself converging simply to id_A. We choose an arbitrary normal state θ on A_p and, for every $i \in I$, define a normal morphism Ψ_i of A_p into itself by putting

$$\Psi_i(pap) = p\Phi_i\ (pap + \theta(pap)(id_A - p))p,$$

$a \in A$; Ψ_i has also finite rank. Moreover (Ψ_i) converges simply to id_{A_p}. □

PROPOSITION 3.10: Let $\{A_i: i \in I\}$ be a family of von Neumann algebras determining $A = \bigoplus_{i\in I} A_i$. Then A is semidiscrete if and only if, for every $i \in I$, A_i is semidiscrete.

PROOF: We denote by u the unit of A. For every $i \in I$, let u_i be the unit of A_i.

1) Let F be a finite subset of I. For every $i \in F$, we choose a normal morphism Φ_i of A_i into itself; we may consider $\bigoplus_{i\in F} \Phi_i \in \mathcal{L}(\bigoplus_{i\in F} A_i)$. Let θ be a normal state on A. We denote by p the projection of A onto $\bigoplus_{i\in F} A_i$ and define $h : \underset{\sim}{C} \to \bigoplus_{i\in I\setminus F} A_i$ by $h(\alpha) = \alpha \sum_{i\in I\setminus F} u_i$ for $\alpha \in \underset{\sim}{C}$. Then

$$\Phi = (\bigoplus_{i\in F} \Phi_i \circ p) + h\circ\theta$$

is a normal morphism of A into A. Moreover Φ has finite rank if and only if, for every $i \in F$, Φ_i has.

Hence if, for every $i \in I$, u_i may be approximated simply by finite-rank normal morphisms of A_i into itself, then u is a simple limit of finite-rank morphisms of A into itself.

2) Let Φ be a normal morphism of A into itself and let again θ be a normal state on A. Fix $i_0 \in I$. For $i \in I$, we put $B_i = A_{i_0}$ and define $\theta_i : B_i \to A_i$ putting, for $a \in B_i$,

$$\theta_i(a) = a \qquad \text{if } i = i_0$$

$$\theta_i(a) = \theta(a)u_i \quad \text{if } i \neq i_0.$$

Let p be the projection of A onto A_{i_0} and let ϕ be the diagonalization morphism of A_{i_0} into $\bigoplus_{i\in I} B_i$ associating to the element a in A_{i_0} the element a in each B_i $(i \in I)$. We consider the normal morphism

$$\psi_{i_0} = p \circ \Phi \circ (\bigoplus_{i\in I} \theta_i) \circ \phi$$

of A_{i_0} into itself. If Φ has finite rank, so has ψ_{i_0}.

Hence if u may be approximated simply by finite-rank normal morphisms mapping A into A, then A_{i_0} may be approximated simply by finite-rank normal morphisms mapping A_{i_0} into A_{i_0}. □

COROLLARY 3.11: Let A and B be von Neumann algebras and let Φ be a normal morphism of A into B. Then A is semidiscrete if and only if B and $\mathcal{Ker}\ \Phi$ are.

PROOF: The statement follows from Proposition 3.10 if one observes that $A \simeq B \oplus \mathcal{Ker}\ \Phi$. □

PROPOSITION 3.12: Let A and B be von Neumann algebras. Then $A \bar{\otimes} B$ is semidiscrete if and only if A and B are.

PROOF: 1) Let Φ [resp. Ψ] be a finite-rank normal morphism of A [resp. B] into itself. There exists a normal morphism Θ of $A \bar{\otimes} B$ into itself such that $\Theta(a \otimes b) = \Phi(a) \otimes \Psi(b)$ whenever $a \in A$, $b \in B$; Θ has finite rank. Put $\Phi_* = {}^t\Phi|_{A_*}$, $\Psi_* = {}^t\Psi|_{B_*}$.

2) Assume the algebras A and B to be semidiscrete.

Let $h_1,\dots,h_m \in (A \bar{\otimes} B)_*$. We want to prove that if $x_1,\dots,x_n \in A \bar{\otimes} B$ and $\varepsilon > 0$, Φ and Ψ may be chosen such that

$$|x_j(h_i) - \Theta(x_j)(h_i)| < \varepsilon$$

whenever $i = 1,\dots,m$ and $j = 1,\dots,n$. We may suppose that, for every

$i = 1,\ldots,m$, $h_i = f_i \otimes g_i$ with $f_i \in A_*$, $g_i \in B_*$.

As the set of all finite-rank normal morphisms of A [resp. B] into itself is convex, from the hypothesis we deduce the existence of a normal morphism Φ [resp. Ψ] of A [resp. B] into itself such that, for all $i = 1,\ldots,m$ and $j = 1,\ldots,n$,

$$\|x_j\| \ \|\Phi_*(f_i) - f_i\| \ \|g_i\| < \varepsilon/2$$

and

$$\|x_j\| \ \|\Psi_*(g_i) - g_i\| \ \|\Phi_* f_i\| < \varepsilon/2.$$

Then also, for Θ associated to Φ and Ψ,

$$|\Theta(x_j)(f_i \otimes g_i) - x_j(f_i \otimes g_i)|$$

$$= |x_j(\Phi_*(f_i) \otimes \Psi_*(g_i) - f_i \otimes g_i)|$$

$$\leq \|x_j\| \ \|\Phi_*(f_i) \otimes \Psi_*(g_i) - f_i \otimes g_i\|$$

$$\leq \|x_j\| \ \|(\Phi_*(f_i)-f_i) \otimes g_i\| + \|x_j\| \ \|\Phi_*(f_i) \otimes (\Psi_*(g_i)-g_i)\|$$

$$\leq \|x_j\| \ \|\Phi_*(f_i)-f_i\| \ \|g_i\| + \|x_j\| \ \|\Phi_*(f_i)\| \ \|\Psi_*(g_i)-g_i\| < \varepsilon.$$

3) Let θ be a normal state on B. One may extend $id_A \otimes \theta$ into a normal morphism F_1 mapping $A \bar{\otimes} B$ into $A \bar{\otimes} \underset{\sim}{C} \simeq A$. Moreover, the canonical injection $F_2 : A \bar{\otimes} \underset{\sim}{C} \to A \bar{\otimes} B$ is a normal morphism. Every normal morphism Φ of $A \bar{\otimes} B$ into itself gives rise to a normal morphism $F_1 \circ \Phi \circ F_2$ of A into itself. Thus if the algebra $A \bar{\otimes} B$ is semidiscrete, so is A. Similarly one establishes semidiscreteness of B. □

PROPOSITION 3.13: If A is a semidiscrete von Neumann algebra, its commutant $A^{\sim}$ constitutes also a semidiscrete von Neumann algebra.

PROOF: Let H be the Hilbert space on which A operates.

1) Suppose that A is a standard von Neumann algebra and let J be the corresponding involution on H. Consider a normal morphism Φ of A into itself. The mapping $x \mapsto Jx^*J$ is a linear antiisomorphism of A onto $A^\sim$ and of $A^\sim$ onto $A^{\sim\sim} = A$; so

$$\Psi : x \mapsto J\Phi(JxJ)J$$

constitutes a normal linear mapping of $A^\sim$ into itself.

Let π be a *-representation of A on a Hilbert space H_1 and let V be an isometry of H into H_1 such that $\Phi = V^{-1}\pi V$. The mapping J induces an involution J_1 on H_1. We have $VJ = J_1V$ and, for $x \in A^\sim$,

$$\Psi(x) = V^{-1}J_1V\Phi(JxJ)V^{-1}J_1V$$

$$= V^{-1}J_1\pi(JxJ)J_1V = JV^{-1}\pi(JxJ)VJ;$$

Ψ constitutes a normal morphism of $A^\sim$ ([T] Chapter IV, Theorem 3.6). Moreover, if Φ admits finite rank, so does Ψ. Finally observe that if $x \in A^\sim$, then

$$(\Psi - \mathrm{id}_{A^\sim})(x) = J((\Phi(JxJ) - \mathrm{id}_A(JxJ))J(x);$$

hence

$$\|\Psi - \mathrm{id}_{A^\sim}\| \leqq \|\Phi - \mathrm{id}_A\|.$$

2) General case

There exists an isomorphism Φ of A onto a standard von Neumann algebra $\Phi(A)$. One may determine a Hilbert space H_2 and a projection p in $(\Phi(A) \otimes \underset{\sim}{C})^\sim = \Phi(A)^\sim \otimes \mathcal{L}(H_2)$ such that A is spatially isomorphic to $p(\Phi(A) \otimes \underset{\sim}{C})p$ ([D] Chapter I, §4, Theorem 3, Corollary). Thus $A^\sim$ is isomorphic to

$$(p(\Phi(A) \otimes \underset{\sim}{C})p)^\sim = p(\Phi(A) \otimes \underset{\sim}{C})^\sim p = p(\Phi(A)^\sim \bar{\otimes} \mathcal{L}(H_2))p.$$

As A is a semidiscrete von Neumann algebra, so is $\Phi(A)$ and, by 1) also $\Phi(A)^\sim$. As $\mathcal{L}(H_2)$ is a semidiscrete von Neumann algebra, so is then

$\Phi(A)^{\sim\sim} \bar{\otimes} \mathcal{L}(H_2)$ by Proposition 3.12; $p(\Phi(A)^{\sim\sim} \bar{\otimes} \mathcal{L}(H_2))p$ is semidiscrete by Proposition 3.9 and finally so is $A^{\sim\sim}$. □

PROPOSITION 3.14: Let A be a von Neumann algebra and let $\{\pi_i : i \in I\}$ be a family of normal morphisms applying A into C*-algebras formed by continuous endomorphisms of Hilbert spaces. The von Neumann algebra A is semidiscrete if and only if, for every $i \in I$, $\pi_i(A)$ is.

PROOF: 1) Assume the condition to hold.

Let $\{p_s : s \in S\}$ be a maximal family of pairwise orthogonal, central projections in A such that, for every $s \in S$, there exists $i = i(s) \in I$ for which $\pi_i|_{p_s A}$ is an isomorphism; the von Neumann algebra $\pi_i(p_s A)$ is semidiscrete. As $A \simeq \bigoplus_{s \in S} \pi_{i(s)}(p_s A)$, semidiscreteness of A follows from Proposition 3.10.

2) The converse statement is established by Corollary 3.11. □

Let A be a von Neumann algebra over a Hilbert space H_1 and let B be a unital C*-algebra that is isomorphic to a *-subalgebra of $\mathcal{L}(H_2)$ for a Hilbert space H_2. Consider a unitary vector ξ in H_1 and a unitary vector η in H_2. We define a linear functional f on $A \otimes_\gamma B$ by putting, for $a \in A$, $b \in B$,

$$f(a \otimes b) = ((a \otimes b)(\xi \otimes \eta) | \xi \otimes \eta)$$

$$= (a\xi \otimes b\eta \mid \xi \otimes \eta) = (a\xi|\xi)_{H_1} (b\eta|\eta)_{H_2};$$

f is a state on $A \otimes_\gamma B$ and we may consider $\Gamma_f(B) \subset A_*$. The family nor of all $f \in S(A \otimes_\gamma B)$ for which $\Gamma_f(B) \subset A_*$ is separating in $S(A \otimes B)$.

If A is a von Neumann algebra and B is a unital C*-algebra, $A \otimes_{nor} B$ is called *normal tensor product* of A and B.

Let A and B be von Neumann algebras. We consider the family bin of all $f \in S(A \otimes B)$ such that the mapping

$$(x,y) \mapsto f(x \otimes y)$$

$$A \times B \to \mathbb{C}$$

is separately weak-*-continuous; bin is also a separating family in $S(A \otimes B)$. We say that $A \otimes_{bin} B$ is the *binormal tensor product* of A and B. We have bin $\subset$ nor and

$$p_{min} \leq p_{bin} \leq p_{nor} \leq p_{max}.$$

<u>LEMMA 3.15</u>: Let A and B be unital C*-algebras. Assume that Γ is a separating family in $S(A \otimes B)$ and, for all $f \in \Gamma$, $y \in A \otimes B$, there exists $g \in \Gamma$ such that

$$f(y^*xy) = f(y^*y)g(x)$$

whenever $x \in A \otimes B$. Then $S_\Gamma(A \otimes B)$ is the weak-*-closure of Γ in $S(A \otimes B) \simeq S_{max}(A \otimes B)$.

<u>PROOF</u>: Let x be a hermitian element in $A \otimes_\Gamma B$ for which $f(x) \geq 0$ whenever $f \in \Gamma$; hence, in particular, $g(x) \geq 0$. We denote by π_f the representation associated to f, over a Hilbert space H_f, for a cyclic vector ξ_f on H_f. For $y \in A \otimes B$,

$$(\pi_f(x)\pi_f(y)\xi_f | \pi_f(y)\xi_f) = f(y^*xy) = f(y^*y)g(x) \geq 0.$$

As $\overline{\pi_f(A \otimes B)\xi_f} = H_f$, we have thus $\pi_f(x) \geq 0$. As Γ is a separating family, $x \geq 0$ follows. Then $S(A \otimes_\Gamma B)$ is the weak-*-closure of Γ (D'] 3.4.1). □

With respect to von Neumann algebras the preceeding lemma applies, in particular, to Γ = min, nor, bin.

<u>LEMMA 3.16</u>: Let A and B be von Neumann algebras. Let F be the convex set of finite-rank, completely normalized mappings from B into A_*, and let Φ_0 be any completely normalized mapping from B into A_*. If $b_1,\ldots,b_n \in B$ and $\varepsilon > 0$, there exists $\Phi \in F$ such that $\Phi(\underset{\sim}{1}) = \Phi_0(\underset{\sim}{1})$ and $\|\Phi(b_i)-\Phi_0(b_i)\| < \varepsilon$ whenever $i = 1,\ldots,n$.

<u>PROOF</u>: Choose $\delta > 0$ such that $\delta^{\frac{1}{2}}(2 + \delta^{\frac{1}{2}}) + \delta < \varepsilon$.

a) As bin is a separating family, bin $\cap$ min is separating in min and, by

Lemma 3.15, bin ∩ min is weak-*-dense in min. Hence the completely normalized mapping Φ_0 may be approximated by elements of F in the weak operator topology. As F is convex, there exists then $\Phi_1 \in F$ such that

$$\|\Phi_1(b_i) - \Phi_0(b_i)\| < \delta/2$$

whenever $i = 1,\dots,n$.

b) We may choose $b_1 = \underset{\sim}{1}$. Put $f = \Phi_0(\underset{\sim}{1})$, $g = \Phi_1(\underset{\sim}{1})$ and $h = g - f$; $\|h\| < \delta/2$. Consider $h_+, h_- \in A_*$; $\|h_+\| + \|h_-\| < \delta/2$.

c) Suppose that $\|b_i\| \leq 1$ whenever $i = 2,\dots,n$. Let θ be a normal state on B. We define $\Psi: B \to A_*$ by putting

$$\Psi(b) = \theta(b)h_-$$

for $b \in B$; Ψ admits rank 1. Let $\Phi_2 = \Phi_1 + \Psi$; Φ_2 has also a finite rank. For every $a \in A$,

$$\Psi(\underset{\sim}{1})\ (a) = \theta(\underset{\sim}{1})h_-(a) = h_-(a).$$

Thus

$$\Phi_2(\underset{\sim}{1}) = \Phi_1(\underset{\sim}{1}) + \Psi(\underset{\sim}{1}) = g + h_- = f + h_+ \geq f. \tag{8}$$

Moreover, for every $b \in B$,

$$\|\Psi(b)\| \leq \|h_-\|\ \|b\| \leq \delta/2\ \|b\|;$$

hence $\|\Psi\| \leq \delta/2 < \delta$ and

$$\|\Phi_2\| < 1 + \delta. \tag{9}$$

For every $i = 1,\dots,n$, we have

$$\begin{aligned}\|\Phi_0(b_i) - \Phi_2(b_i)\| &\leq \|\Phi_1(b_i) - \Phi_0(b_i)\| + \|\Psi(b_i)\| \\ &< \delta/2 + \delta/2 = \delta.\end{aligned}$$

d) Let $k = \Phi_2(\underset{\sim}{1})$ and consider the representation π, associated to the positive functional k, over a Hilbert space H for a cyclic vector ξ. As $k \geqq f$, by (8), there exists $c \in A$ such that $0 \leqq c \leqq \underset{\sim}{1}$ and, for every $a \in A$,

$$f(a) = k(cac) = (\pi(cac)\xi|\xi) = (\pi(a)\pi(c)\xi|\pi(c)\xi) \qquad (10)$$

([D'] 2.5.1; [S] Theorem 1.24.3). Put $\eta = \pi(c)\xi \in H$. For every $a \in A$,

$$f(a) = (\pi(a)\eta|\eta).$$

In particular,

$$f(\underset{\sim}{1}) = (\eta|\eta) = \|\eta\|^2.$$

As $f(\underset{\sim}{1}) = \underset{\sim}{1}$, $\|\eta\| = 1$. As $0 \leqq c \leqq \underset{\sim}{1}$, we have $c^2 \leqq c$, hence

$$(\eta|\xi) = (\pi(c)\xi|\xi) \geqq (\pi(c^2)\xi|\xi) = \|\pi(c)\xi\|^2 = \|\eta\|^2 = 1.$$

Taking relation (9) into account we obtain

$$\|\xi-\eta\|^2 = \|\xi\|^2 - 2(\eta|\xi) + \|\eta\|^2$$

$$= k(\underset{\sim}{1}) - 1 \leqq \|\Phi_2\| - 1 < \delta,$$

$$\|\xi-\eta\| < \delta^{1/2}$$

and

$$\|\xi\| \leqq \|\xi-\eta\| + \|\eta\| < 1 + \delta^{1/2}.$$

We finally define $\Phi: B \to A_*$ by putting

$$\Phi(b)(a) = \Phi_2(b)(cac)$$

for $a \in A$, $b \in B$; Φ has also finite rank and, by (10),

$$\Phi(\underset{\sim}{1})(a) = \Phi_2(\underset{\sim}{1})(cac) = k(cac) = f(a).$$

Thus $\Phi(\underset{\sim}{1}) = f = \Phi_0(\underset{\sim}{1})$.

We may determine a positive linear functional ϕ on $A \otimes_{bin} B$ such that $\Gamma_\phi = \Phi_2$. Define

$$\psi(x) = \phi((c \otimes \underset{\sim}{1})\ x\ (c \otimes \underset{\sim}{1}))$$

for $x \in A \otimes_{bin} B$; ψ is a state and $\Phi = \Gamma_\psi \in F$. For $i = 1,\ldots,n$, let

$$b_i' = \theta_k^{-1}(\Phi_2(b_i)) \in \pi(A)^{\sim\sim}.$$

For every $a \in A$,

$$\langle a, \theta_k(\underset{\sim}{1})\rangle = (\pi_a \xi | \xi) = k(a);$$

hence $\theta_k(\underset{\sim}{1}) = k$ and $\underset{\sim}{1} = \theta_k^{-1}(k) = \theta_k^{-1}(\Phi_2(\underset{\sim}{1}))$. Therefore

$$\|b_i'\| \leq \|b_i\| \quad (\leq 1)$$

whenever $i = 1,\ldots,n$. Moreover, as $\|k\| \leq 1$, we have also $\|\pi\| \leq 1$. For every $a \in A$ such that $\|a\| \leq 1$, we obtain

$$|(\Phi(b_i) - \Phi_2(b_i))(a)| = |\Phi_2(b_i)(cac-a)|$$

$$= |\theta_k(b_i')(cac-a)| = |(\pi(cac-a)b_i'\xi|\xi)|$$

$$= |(b_i'\pi(cac-a)\xi|\xi)|$$

$$= |(b_i'\pi(a)\pi(c)\xi|\pi(c)\xi) - (b_i'\pi(a)\xi|\xi)|$$

$$\leq |(b_i'\pi(a)\eta|\eta - \xi)| + |(b_i'\pi(a)(\eta-\xi)|\xi)|$$

$$\leq \|\eta-\xi\| \ (1 + \|\xi\|)$$

$$\leq \delta^{1/2}\ (2 + \delta^{1/2}) < \varepsilon - \delta.$$

Thus

$$\|\Phi(b_i) - \Phi_2(b_i)\| < \varepsilon - \delta$$

and

$$\|\Phi(b_i)-\Phi_0(b_i)\| \leqq \|\Phi(b_i)-\Phi_2(b_i)\| + \|\Phi_2(b_i)-\Phi_0(b_i)\|$$

$$< \varepsilon - \delta + \delta = \varepsilon. \quad \square$$

<u>THEOREM 3.17</u>: Let A be a von Neumann algebra. The following properties are equivalent:

(i) The algebra A is semidiscrete.

(ii) For every von Neumann algebra B, one has

$$A \otimes_{bin} B = A \otimes_{min} B, \text{ i.e., } p_{bin} = p_{min} \text{ on } A \otimes B.$$

(ii') For every von Neumann algebra A_1 admitting A as a subalgebra and every von Neumann algebra B, one has $A \otimes_{bin} B \subset A_1 \otimes_{bin} B$, i.e., the canonical injection of $(A \otimes B, p_{bin})$ into $(A_1 \otimes B, p_{bin})$ is an isometry.

(ii") For every von Neumann algebra B_1 and every von Neumann subalgebra B of B_1, one has $A \otimes_{bin} B \subset A \otimes_{bin} B_1$, i.e., the canonical injection of $(A \otimes B, p_{bin})$ into $(A \otimes B_1, p_{bin})$ is an isometry.

(iii) For every C*-algebra B, one has $A \otimes_{nor} B = A \otimes_{min} B$, i.e., $p_{nor} = p_{min}$ on $A \otimes B$.

(iii') For every von Neumann algebra A_1 admitting A as a subalgebra and every C*-algebra B, one has $A \otimes_{nor} B \subset A_1 \otimes_{nor} B$, i.e., the canonical injection of $(A \otimes B, p_{nor})$ into $(A_1 \otimes B, p_{nor})$ is an isometry.

<u>PROOF</u>: (i) => (iii)

Recall that $p_{min} \leqq p_{nor}$; min $\subset$ nor.

Let $f \in$ nor; hence $\Gamma_f(B) \subset A_*$. As A is semidiscrete, there exists a net (Φ_i) of finite-rank normal morphisms mapping A_* into itself such that $\lim_i \|\Phi_i(\phi) - \phi\| = 0$ whenever $\phi \in A_*$. The net $(\Phi_i \circ \Gamma_f)$ consists of finite-rank completely normalized mappings and converges to Γ_f in the weak operator topology. Hence f is a weak-*-limit of states in min. Therefore, by Lemma 3.15, we have min = nor.

(iii) => (iii') [resp. (ii) => (ii')]

By hypothesis there exists an isometric injection ψ of $A \otimes_{nor} B$ [resp. $A \otimes_{bin} B$] into $A \otimes_{min} B$. Moreover, there exists an isometric injection ϕ' of $A \otimes_{min} B$ into $A_1 \otimes_{min} B$. Denote by ϕ the canonical continuous injection of $A \otimes_{nor} B$ [resp. $A \otimes_{bin} B$] into $A_1 \otimes_{nor} B$ [resp. $A_1 \otimes_{bin} B$] and by ψ' the canonical continuous injection of $A_1 \otimes_{nor} B$ [resp. $A_1 \otimes_{bin} B$] into $A_1 \otimes_{min} B$. Hence $\psi' \circ \phi = \phi' \circ \psi$ is an isometry. For every $x \in A \otimes_{nor} B$ [resp. $x \in A \otimes_{bin} B$], $\|x\| = \|\psi'\circ\phi(x)\| \leq \|\phi(x)\| \leq \|x\|$. Thus ϕ is an isometry.

(iii) => (ii)

As $p_{min} \leq p_{bin} \leq p_{nor}$, the hypothesis implies that $A \otimes_{bin} B = A \otimes_{min} B$.

(ii) => (ii")

As $A \otimes_{min} B \subset A \otimes_{min} B_1$, the hypothesis implies that $A \otimes_{bin} B \subset A \otimes_{bin} B_1$.

(iii') => (iii) [resp. (ii') => (ii)]

Let H be the Hilbert space on which A operates. By hypothesis the canonical injection ϕ of $A \otimes_{nor} B$ [resp. $A \otimes_{bin} B$] into $\mathcal{L}(H) \otimes_{nor} B$ [resp. $\mathcal{L}(H) \otimes_{bin} B$] is isometric. As $\mathcal{L}(H)$ is a semidiscrete von Neumann algebra and (i) => (iii) [resp. (i) => (iii) => (ii)], the canonical bijection ψ of $\mathcal{L}(H) \otimes_{nor} B$ [resp. $\mathcal{L}(H) \otimes_{bin} B$] onto $\mathcal{L}(H) \otimes_{min} B$ is isometric. Moreover, consider the canonical continuous injection ϕ' of $A \otimes_{nor} B$ [resp. $A \otimes_{bin} B$] into $A \otimes_{min} B$ and the canonical continuous injection ψ' of $A \otimes_{min} B$ into $\mathcal{L}(H) \otimes_{min} B$; $\psi' \circ \phi' = \psi \circ \phi$ is an isometry. For every $x \in A \otimes_{nor} B$ [resp. $x \in A \otimes_{bin} B$], $\|x\| = \|\psi' \circ \phi'(x)\| \leq \|\phi'(x)\| \leq \|x\|$; therefore ϕ' is an isometry.

(ii") => (ii)

Let H be the Hilbert space on which B operates. Denote by ϕ the canonical continuous injection of $A \otimes_{min} B$ into $A \otimes_{min} \mathcal{L}(H)$ and by ϕ' the canonical continuous injection of $A \otimes_{bin} B$ into $A \otimes_{min} B$. There exists an isometric injection ψ of $A \otimes_{bin} B$ into $A \otimes_{bin} \mathcal{L}(H)$. As $\mathcal{L}(H)$ is a semidiscrete von Neumann algebra and (i) => (iii) => (ii), there exists also an isometric injection ψ' of $A \otimes_{bin} \mathcal{L}(H)$ onto $A \otimes_{min} \mathcal{L}(H)$. Thus $\phi \circ \phi' = \psi' \circ \psi$ is an isometry. For every $x \in A \otimes_{bin} B$,

$$\|x\| = \|\phi \circ \phi'(x)\| \leqq \|\phi'(x)\| \leqq \|x\|;$$

hence ϕ' is an isometry.

(ii) => (i)

Let f be a normal state on A and let π be the associated *-representation on a Hilbert space H for the cyclic vector ξ; then $\theta_f : \pi(A)^{\sim} \to A_*$. By Lemma 3.16 applied to the von Neumann algebra $\pi(A)^{\sim}$, there exists a net $(\Phi_i)_{i\in I}$ of morphisms mapping $\pi(A)^{\sim}$ into A_*, converging to θ_f in the weak operator topology, admitting finite rank and for which $\Phi_i(\underset{\sim}{1}) = \theta_f(\underset{\sim}{1}) = f$ whenever $i \in I$.

Let $i \in I$ be fixed and put $\Psi_i = \theta_f^{-1} \circ \Phi_i$; Ψ_i is a finite rank morphism of $\pi(A)^{\sim}$ into itself. We choose a net $(s_{i'})_{i'\in I'}$ in $\pi(A)^{\sim}$ such that $\|s_{i'}\| \leqq 1$ whenever $i' \in I'$ and $(s_{i'})$ is weakly convergent to 0. If $a_1, a_2 \in A$ and $i' \in I'$, we have

$$(\Psi_i(s_{i'})\pi(a_1)\xi|\pi(a_2)\xi) = (\pi(a_2^*)\theta_f^{-1}(\Phi_i(s_{i'}))\pi(a_1)\xi|\xi)$$

$$= (\theta_f^{-1}(\Phi_i(s_{i'}))\pi(a_2^*)\pi(a_1)\xi|\xi)$$

$$= (\theta_f^{-1}(\Phi_i(s_{i'}))\pi(a_2^*a_1)\xi|\xi)$$

$$= \Phi_i(s_{i'})(a_2^*a_1).$$

Hence

$$\lim_{i'} (\Psi_i(s_{i'})\,\pi(a_1)\xi|\pi(a_2)\xi) = 0$$

and, as ξ is a cyclic vector, $(\Psi_i(s_{i'}))_{i'}$ converges weakly to 0; Ψ_i is normal.

Let $b \in \pi(A)^{\sim}$. For all $a_1, a_2 \in A$ and $i \in I$, we have also

$$((\Psi_i(b)-b)\pi(a_1)\xi|\pi(a_2)\xi)$$

$$= ((\theta_f^{-1} \circ \Phi_i(b)-b)\pi(a_1)\xi|\pi(a_2)\xi)$$

$$= (\pi(a_2^*)\theta_f^{-1} \circ (\Phi_i-\theta_f)(b)\pi(a_1)\xi|\xi)$$

$$= (\theta_f^{-1} \circ (\Phi_i-\theta_f)(b)\pi(a_2^* a_1)\xi|\xi)$$

$$= (\Phi_i-\theta_f)(b)(a_2^* a_1).$$

Thus

$$\lim_i ((\Psi_i(b)-b)\pi(a_1)\xi|\pi(a_2)\xi) = 0.$$

As ξ is a cyclic vector for π, we conclude that $(\Psi_i(b)-b)$ converges weakly to 0.

Hence (Ψ_i) converges to $id_{\pi(A)^\sim}$ in the weak operator topology. Therefore $\pi(A)^\sim$ is a semidiscrete von Neumann algebra. By Proposition 3.13, the von Neumann algebra $\pi(A) = \pi(A)^{\sim\sim}$ is also semidiscrete. From Proposition 3.14 we deduce then the semidiscreteness of the von Neumann algebra A. □

Let A be a von Neumann algebra over the Hilbert space H. Let ξ be a unitary vector in H and consider the associated normal state ω_ξ on A. If $a \in A$ and $b \in A^\sim$, we have

$$\omega_\xi \circ \bar{\omega}(a \otimes b) = \omega_\xi(ab) = (ab\xi|\xi);$$

so we define $\omega_\xi \circ \bar{\omega} \in S_{bin}(A \otimes A^\sim)$. For all $a \in A$, $c \in A$ and $b \in A^\sim$, $d \in A^\sim$, hence $b^* \in A^\sim$, $d^* \in A^\sim$, we have

$$\bar{\omega}((a \otimes b))^* \bar{\omega}(c \otimes d) = (ab)^* (cd) = b^*a^*cd = a^*cb^*d$$

$$= \bar{\omega}(a^*c \otimes b^*d) = \bar{\omega}((a \otimes b)^* c \otimes d);$$

thus, for every $x \in A \otimes A^\sim$, $\bar{\omega}(x)^*\bar{\omega}(x) = \bar{\omega}(x) = \bar{\omega}(x^*x)$. Moreover, for every $X \in A \otimes A^\sim$,

$$\|\bar{\omega}(x)\xi\|^2 = (\bar{\omega}(x)\xi|\bar{\omega}(x)\xi) = (\bar{\omega}(x)^* \bar{\omega}(x)\xi|\xi)$$

$$= \omega_\xi(\bar{\omega}(x)^* \bar{\omega}(x)) = \omega_\xi(\bar{\omega}(x^*x));$$

hence

$$\|\bar{\omega}(x)\| = \sup\{\|\bar{\omega}(x)\xi\| : \xi \in H,\ \|\xi\| = 1\}$$

$$= \sup\ \{\omega_\xi \circ \bar{\omega}(x^*x)^{1/2} : \xi \in H,\ \|\xi\| = 1\} \leqq \|x\|_{bin}.$$

Therefore $\bar{\omega}$ admits an extension to a continuous homomorphism of $A \otimes_{bin} A^\sim$ into $\mathcal{L}(H)$.

THEOREM 3.18: Let A be a von Neumann algebra over the Hilbert space H. Then the homomorphism $\bar{\omega}$ may be extended continuously on $A \otimes_{min} A^\sim$ if and only if A is semidiscrete.

PROOF: 1) If A is semidiscrete, so is $A \bar{\otimes} A^\sim$ by Propositions 3.13 and 3.12. We deduce from Theorem 3.17 that, for every $x \in A \bar{\otimes} A^\sim$,

$$\|\bar{\omega}(x)\| \leqq \|x\|_{bin} = \|x\|_{min}.$$

2) Assume the extension to exist and denote it also by $\bar{\omega}$. For every unitary vector ξ in H,

$$\omega_\xi \circ \bar{\omega}(x) = (\bar{\omega}(x)\xi|\xi),$$

$x \in A \otimes A^\sim$; $\omega_\xi \circ \bar{\omega} \in S_{bin}(A \otimes A^\sim)$. By hypothesis we have also $\omega_\xi \circ \bar{\omega} \in S_{min}(A \otimes A^\sim)$. We denote by T the completely normalized mapping of $A^\sim$ into A_* that is associated to $\omega_\xi \circ \bar{\omega}$; we put $f = T(\underset{\sim}{1})$, $\underset{\sim}{1}$ denoting the unit element in $A^\sim$. Let π be the representation associated to f. For every $a \in A$,

$$f(a) = T(\underset{\sim}{1})(a) = \omega_\xi \circ \bar{\omega}(a \otimes \underset{\sim}{1}) = (a\xi|\xi). \tag{11}$$

Let p be the projection of $\mathcal{L}(H)$ onto $\overline{A\xi}$; $p \in A^\sim$. Via (11) we may identify $pA = Ap$ with $\pi(A)$, and then also $pA^\sim p = (pAp)^\sim$ with $\pi(A)^\sim$. We now consider the morphism $\theta_f^{-1} \circ T : A^\sim \to \pi(A)^\sim$. If $b \in A^\sim$, $b' = \theta_f^{-1} \circ T(b)$ is the unique element in $pA^\sim p$ for which

$$T(b)(a) = (ab'\xi|\xi)$$

whenever $a \in A$. But, for every $a \in A$,

$$T(b)(a) = \omega_\xi \circ \bar{\omega}(a \otimes b) = (ab\xi|\xi) = (appb\xi|\xi) = (apbp\xi|\xi).$$

Hence

$$\theta_f^{-1} \circ T(b) = pbp. \tag{12}$$

By Lemma 3.16 there exists a net $(T_i)_{i\in I}$ of finite-rank completely normalized mappings from $A^{\sim\sim}$ into A_* converging to T in the weak operator topology. Thus the net $(\theta_f^{-1} \circ T_i)$ consisting of finite-rank morphisms mapping $A^{\sim\sim}$ into $pA^{\sim\sim}p$ converges to $\theta_f^{-1} \circ T$ with respect to the weak operator topology.

Let θ be a normal state on $A^{\sim\sim}_p$. If $i \in I$, we define a normal morphism Φ_i of $A^{\sim\sim}_p$ into itself putting

$$\Phi_i(x) = \theta_f^{-1} \circ T_i(x + \theta(x)\ (\underset{\sim}{1}-p))$$

for $x \in A^{\sim\sim}_p$; Φ_i has finite rank. The relation (12) implies convergence of (Φ_i) to $id_{A^{\sim\sim}_p}$ with respect to the weak operator topology. Thus the von Neumann algebra $A^{\sim\sim}_p$ is semidiscrete. As $pAp = pA^{\sim\sim\sim\sim}p = (pA^{\sim\sim}p)^{\sim\sim}$, semidiscreteness of pAp follows from Proposition 3.13. From Proposition 3.14 we conclude finally that the von Neumann algebra A is semidiscrete. □

C. Relations between remarkable properties of von Neumann algebras

The semidiscrete von Neumann algebras are precisely the amenable von Neumann algebras. In view of establishing that important fact, we consider some other significant properties of von Neumann algebras and show their equivalence with amenability.

PROPOSITION 3.19: Every semidiscrete von Neumann algebra A is nuclear.

PROOF: Let B be a C*-algebra. We consider the canonical homomorphism Φ of

$A \otimes_{max} B$ onto $A \otimes_{min} B$ and show that every state, defined on $A \otimes_{max} B_1$ vanishes on $\mathit{Ker}\ \Phi$.

Let f be a state on $A \otimes_{max} B$ for which Γ_f has finite rank. Thus there exist $g_1,\dots,g_n \in A^*$ and $h_1,\dots,h_n \in B^*$ such that $\Gamma_f(b)(a) = \sum_{i=1}^{n} g_i(a)h_i(b)$ whenever $a \in A$, $b \in B$. Therefore $f = \sum_{i=1}^{n} g_i \otimes h_i$ is continuous on $A \otimes_{min} B$ and $\mathit{Ker}\ \Phi \subset \mathit{Ker}\ f$.

Semidiscreteness of A now implies that, for every state f on $A \otimes_{max} B$, $\mathit{Ker}\ \Phi \subset \mathit{Ker}\ f$. □

DEFINITION 3.20: The von Neumann algebra A over the Hilbert space H is called injective if there exists a projection of norm 1 in $\mathcal{L}(\mathcal{L}(H))$ mapping $\mathcal{L}(H)$ onto A.

Any injective von Neumann algebra is generated by an increasing net of injective von Neumann algebras that are countably generated ([25] Theorem 4). Every crossed product of an injective von Neumann algebra by the group $\mathbb{R}$ is injective ([17] Proposition 6.8).

LEMMA 3.21: Every von Neumann algebra A over a Hilbert space H, for which every completely normalized mapping of A into A may be extended to a completely normalized mapping of $\mathcal{L}(H)$ into A, is injective.

PROOF: One considers the completely normalized mapping id_A of A into A and one observes that its extension to a completely normalized mapping of $\mathcal{L}(H)$ into A is necessarily a projection of norm 1. □

PROPOSITION 3.22 If A is a nuclear C*-algebra, its enveloping von Neumann algebra A** is injective.

PROOF: Let B_1 be a C*-algebra and let B be a C*-subalgebra of B_1.. We consider $A \otimes B$ to be a subalgebra of $A \otimes B_1$; $A \otimes_{min} B \subset A \otimes_{min} B_1$. By hypothesis $A \otimes_{max} B = A \otimes_{min} B$ and $A \otimes_{max} B_1 = A \otimes_{min} B_1$. Thus the restriction of $S_{max}(A \otimes B_1)$ to $S_{max}(A \otimes B)$ is surjective; then, for the von Neumann algebra A**, the restriction of nor$(A^{**} \otimes B_1)$ to nor $(A^{**} \otimes B)$ is also surjective. Every completely normalized mapping of B into A** may be extended to a completely normalized mapping of B_1 into A**. Injectivity of the von Neumann algebra A^{**} follows now from Lemma 3.21 applied to $B = A^{**}$, $B_1 = \mathcal{L}(H)$, H being the Hilbert space on which A^{**} operates. □

COROLLARY 3.23: Every nuclear von Neumann algebra is injective.

PROOF: The statement is an immediate consequence of Proposition 3.22. □

We quote an important technical result. Let A be a C*-algebra. If Φ is a bounded bilinear functional on $A \times A$, we may determine states f_1, f_2, g_1, g_2 on A such that the following *generalized Grothendieck inequality*

$$|\Phi(a,b)| \leq \sup \{|\Phi(x,y)|: x \in A, y \in A, \|x\| \leq 1, \|y\| \leq 1\}$$

$$(f_1(a^*a) + f_2(a\,a^*))^{1/2}(g_1(b^*b) + g_2(bb^*))^{1/2}$$

holds for all $a,b \in A$. In case A is a von Neumann algebra and Φ is separately ultraweakly continuous, the states may be chosen to be normal ([37] Theorem 1.1; [49]).

If A is a von Neumann algebra over the Hilbert space H, for every $a \in I(A)$, $a^*a = id_H$ and the projection aa^* is majorized by id_H. Hence in case A is finite, $aa^* = id_H$ and we have $I(A) = A_u$.

We establish a collection of lemmas in which we consider a mean μ on $\ell^\infty(I(A))$ for a von Neumann algebra A. If $f \in \ell^\infty(I(A))$, we denote $\langle f,\mu\rangle$ by $\int_{I(A)} f(v)d\mu(v)$.

LEMMA 3.24: Let A be a von Neumann algebra and let p be a maximal central projection for which the von Neumann algebra A_p is finite. We consider a mean μ on $\ell^\infty(I(A))$ such that

$$\int_{I(A)} f(vv^*)d\mu(v) = f(p)$$

for every normal positive functional f on A. Then for every bounded separately ultraweakly continuous bilinear functional Φ on $A \times A$, the functionals

$$M : x \mapsto \int_{I(A)} \Phi(xv^*,v)d\mu(v)$$

$$N : x \mapsto \int_{I(A)} \Phi(v^*,vx)d\mu(v)$$

on A are normal.

PROOF: Let f_1, f_2, g_1, g_2 be the states associated to Φ via the generalized Grothendieck inequality. For i = 1,2, we put

$$f_i'(x) = \int_{I(A)} f_i(vxv^*)d\mu(v) = \int_{I(A)} v^*f_iv(x)d\mu(v),$$

$$g_i'(x) = \int_{I(A)} g_i(vxv^*)d\mu(v) = \int_{I(A)} v^*g_iv(x)d\mu(v),$$

$x \in A$. If ϕ is any normal positive functional on the finite von Neumann algebra A_p, then the convex hull of $\{v^*\phi v: v \in (A_p)_u\}$ is relatively compact in the weak topology of $(A_p)_*$ ([T] Chapter V; proof of 2.4). Therefore the restrictions of f_i', g_i' (i = 1,2) to A_p are normal functionals. As p belongs to the center of A, so does $\underset{\sim}{1}$-p, $\underset{\sim}{1}$ being the unit element in A. For i = 1,2,

$$f_i'(\underset{\sim}{1}-p) = \int_{I(A)} f_i(v(\underset{\sim}{1}-p)v^*)d\mu(v)$$

$$= \int_{I(A)} f_i(vv^*(\underset{\sim}{1}-p))d\mu(v) = \int_{I(A)} ((\underset{\sim}{1}-p)f_i)(vv^*)d\mu(v);$$

by hypothesis the latter expression is $(\underset{\sim}{1}-p)f_i(p) = f_i(p(\underset{\sim}{1}-p)) = 0$. Thus f_1' $f_2' \in A_*$; similarly g_1', $g_2' \in A_*$.

Let $\alpha = \sup\{|\Phi(x,y)|: x \in A,\ y \in A,\ \|x\| \leq 1,\ \|y\| \leq 1\}$. If $v \in I(A)$ and $a \in A$, we have

$$|\Phi(av^*,v)| \leq \alpha(f_1(va^*av^*) + f_2(aa^*))^{\frac{1}{2}} (g_1(\underset{\sim}{1}) + g_2(\underset{\sim}{1}))^{\frac{1}{2}}$$

$$= \alpha\sqrt{2}\ (f_1(va^*av^*) + f_2(aa^*))^{\frac{1}{2}}.$$

As μ is a mean, Hölder's inequality implies that

$$|M(a)| \leqq \alpha\sqrt{2}\,(f_1'(a^*a) + f_2(aa^*))^{\frac{1}{2}};$$

similarly

$$|N(a)| \leqq \alpha\sqrt{2}\,(g_1(a^*a) + g_2'(aa^*))^{\frac{1}{2}}.$$

Hence, $M, N \in A_*$. □

LEMMA 3.25: Let $(A_i)_{i\in I}$ be an increasing net of von Neumann algebras over the same Hilbert space and let A be the ultraweak closure of $\bigcup_{i\in I} A_i$. Suppose that, for every $i \in I$, there exists a mean μ_i on $\ell^\infty(I(A_i))$ such that

$$\int_{I(A_i)} \Phi_i(av^*, v)\,d\mu_i(v) = \int_{I(A_i)} \Phi_i(v^*, va)\,d\mu_i(v)$$

whenever Φ_i is a bounded, separately ultraweakly continuous, bilinear functional on $A_i \times A_i$ and $a \in A_i$. Then there exists a mean σ on $\ell^\infty(I(A))$ such that

$$\int_{I(A)} \Phi(av^*, v)\,d\sigma(v) = \int_{I(A)} \Phi(v^*, va)\,d\sigma(v)$$

for every bounded, separately ultraweakly continuous bilinear functional Φ on $A \times A$, and every $a \in A$. If p is the maximal finite central projection in A, then

$$\int_{I(A)} F(vv^*)\,d\sigma(v) = F(p)$$

whenever F is a normal positive functional on A.

PROOF: For every $i \in I$, $\Phi|_{A_i \times A_i}$ is a separately ultraweakly continuous, bilinear functional on $A_i \times A_i$ and

$$\int_{I(A_i)} \Phi(av^*, v)\,d\mu_i(v) = \int_{I(A_i)} \Phi(v^*, va)\,d\mu_i(v)$$

whenever $a \in A_i$. One may extend μ_i to a mean ν_i on $\ell^\infty(I(A))$ putting

$$\nu_i(f) = \mu_i(f|_{A_i})$$

for $f \in \ell^\infty(I(A))$. If $i,j \in I$, $i < j$ and $a \in A_i$, we have

$$\int_{I(A)} \Phi(av^*,v)d\nu_j(v) = \int_{I(A)} \Phi(v^*,va)d\nu_j(v).$$

Let μ be a weak-*-limit point of (ν_i) in the weak-*-compact set $\mathfrak{m}(\ell^\infty(I(A)))$. Then also

$$\int_{I(A)} \Phi(av^*,v)d\mu(v) = \int_{I(A)} \Phi(v^*,va)d\mu(v)$$

whenever $a \in \bigcup_{i\in I} A_i$.

1) Suppose A to be a finite von Neumann algebra.

The maximal finite central projection is id_A. For every normal positive functional F on A we have trivially

$$\int_{I(A)} F(vv^*)d\mu(v) = \int_{I(A)} F(id_A)d\mu(v) = F(id_A).$$

The first statement follows from Lemma 3.24 with $\sigma = \mu$.

2) Suppose A to be infinite and let p be the maximal finite central projection. If $\underset{\sim}{1}$ denotes the unit element in A, $\underset{\sim}{1}$-p is also central; $\underset{\sim}{1}$-p is properly infinite ([T] Chapter V, Theorem 1.19). The von Neumann algebra $A_{\underset{\sim}{1}-p}$ is properly infinite; it contains a type I factor of the form $\mathcal{L}(H_o)$ for an infinite-dimensional separable Hilbert space H_o. In $\mathcal{L}(H_o)$ one may determine an infinite increasing sequence (p_n) of central projections and a sequence (w_n) of partially isometric operators such that, for every $n \in \underset{\sim}{N}^*$, $w_n^* w_n = p_n$ and $w_n w_n^* \leq p_n$. For $n \in \underset{\sim}{N}^*$, let $H_n = w_n H_o$; we have

$$H_n^\perp = \{\eta \in H_o : (\forall\xi \in H_o)\ (w_n\xi|\eta) = 0\}$$

$$= \{\eta \in H_o : (\forall\xi \in H_o)\ (\xi|w_n^*\eta) = 0\} = Ker\ w_n^*.$$

As (H_n) is an increasing sequence and $H_o = \bigcup_n H_n$, $(Ker\ w_n^*)$ is a decreasing sequence for which $\bigcap_n Ker\ w_n^* = \{0\}$. Thus (w_n^*) converges ultrastrongly to 0 and $(w_n w_n^*)$ converges ultraweakly to 0. For $n \in \tilde{N}^*$, let $v_n = p + w_n$; as p [resp. w_n] is a partial isometry of pA [resp. $(\tilde{1}-p)A$], we have $v_n \in I(A)$. Moreover, for every $n \in \tilde{N}^*$, $v_n v_n^* = p + w_n w_n^*$; $(v_n v_n^*)$ converges ultraweakly to p. If $n \in \tilde{N}^*$, we define a mean σ_n on $\ell^\infty(I(A))$ putting

$$\langle f,\sigma_n\rangle = \int_{I(A)} f(v_n v)d\mu(v)$$

for $f \in \ell^\infty(I(A))$. Let σ be a weak-*-limit point of (σ_n) in $\mathfrak{m}(\ell^\infty(I(A)))$.

Let Φ be a bounded, separately ultraweakly continuous, bilinear functional on $A \times A$. If $n \in \tilde{N}^*$, we define another functional Φ_n of that form putting

$$\Phi_n(a,b) = \Phi(av_n^*,v_n b)$$

for $a,b \in A$.

Let $a \in \bigcup_{i\in I} A_i$. If $n \in \tilde{N}^*$ and $v \in I(A)$, we have

$$\Phi_n(av^*,v) = \Phi(av^*v_n^*,v_n v),$$

$$\Phi_n(v^*,va) = \Phi(v^*v_n^*,v_n va);$$

$$\int_{I(A)} \Phi(av^*,v)d\sigma_n(v) = \int_{I(A)} \Phi(av^*v_n^*,v_n v)d\mu(v)$$

$$= \int_{I(A)} \Phi_n(av^*,v)d\mu(v) = \int_{I(A)} \Phi_n(v^*,va)d\mu(v)$$

$$= \int_{I(A)} \Phi(v^*v_n^*,v_n va)d\mu(v) = \int_{I(A)} \Phi(v^*,va)d\sigma_n(v).$$

With respect to the initial considerations we obtain

$$\int_{I(A)} \Phi(av^*,v)d\sigma(v) = \int_{I(A)} \Phi(v^*,va)d\sigma(v).$$

Let F be a normal positive functional on A. For every $v \in I(A)$, $\|v\| = 1$ and $p \leq vv^*$. As σ is a mean, we have then

$$F(p) \leq \int_{I(A)} F(vv^*)d\sigma(v).$$

By the definition of σ,

$$\int_{I(A)} F(vv^*)d\sigma(v) \leqq \limsup_{n \to \infty} \int_{I(A)} F(v_n vv^* v_n^*)d\mu(v).$$

But $vv^* \leqq \underset{\sim}{1}$; hence, for every $n \in \underset{\sim}{N}^*$, $v_n vv^* v_n^* \leqq v_n v_n^*$ and

$$\limsup_{n \to \infty} \int_{I(A)} F(v_n vv^* v_n^*)d\mu(v)$$

$$\leqq \limsup_{n \to \infty} \int_{I(A)} F(v_n v_n^*)d\mu(v) = \limsup_{n \to \infty} F(v_n v_n^*).$$

As $(v_n v_n^*)$ converges ultraweakly to p and F is a normal state, we have

$$\limsup_{n \to \infty} F(v_n v_n^*) = F(p).$$

Thanks to Lemma 3.24, the proposition is established for the mean μ. □

PROPOSITION 3.26: Let A be an injective von Neumann algebra. There exists a mean μ on $\ell^\infty(I(A))$ such that

$$\int_{I(A)} \Phi(av^*,v)d\mu(v) = \int_{I(A)} \Phi(v^*,va)d\mu(v)$$

whenever Φ is a bounded, separately ultraweakly continuous, bilinear functional on $A \times A$, and $a \in A$.

PROOF: 1) Suppose A to be finite-dimensional.

Every bilinear functional on $A \times A$ that is separately ultraweakly continuous is also separately continuous. Via the Hahn-Banach theorem we may extend the normalized Haar integral on the compact group A_u to a mean μ on $\ell^\infty(A_u)$. As the mean μ is right invariant on $C(A_\mu)$, for every $a \in A_u$, we have

$$\int_{A_u} \Phi(av^*,v)d\mu(v) = \int_{A_u} \Phi(a(va)^*,va)d\mu(v)$$

$$= \int_{A_u} \Phi(v^*,va)d\mu(v).$$

As A_u generates A and $A_u = I(A)$ in the finite-dimensional algebra A, the proposition is established in that case.

2) By Lemma 3.25 the proposition then holds for any injective von Neumann algebra with separable predual. One may carry the result over to any countably generated injective von Neumann algebra because such an algebra is the direct sum of injective von Neumann algebras with separable preduals. Finally the statement is generalized for all injective von Neumann algebras. □

THEOREM 3.27: Every nuclear C*-algebra is amenable.

PROOF: We identify $(A \hat{\otimes} A)^*$ with the space $\mathcal{L}_2(A)$ of continuous bilinear functionals on $A \times A$ by putting

$$\langle a \otimes b, \Phi\rangle = \Phi(a,b)$$

for $\Phi \in \mathcal{L}_2(A)$ and $a,b \in A$. Every $\Phi \in \mathcal{L}_2(A)$ admits an extension Φ_1 to a separately ultraweakly continuous bilinear functional for the enveloping von Neumann algebra A^{**}. The latter is injective by Proposition 3.22; we associate to A^{**} a mean μ on $\ell^\infty(I(A^{**}))$ satisfying the condition of Proposition 3.26. We put

$$\Psi(\Phi) = \int_{I(A^{**})} \Phi_1(v^*,v)d\mu(v),$$

$\Phi \in \mathcal{L}_2(A)$. Then $\Psi \in \mathcal{L}_2(A)^* = (A \hat{\otimes} A)^{**}$. We show that Ψ constitutes a virtual diagonal for A.

For all $\Phi \in \mathcal{L}_2(A)$, $a \in A$, by Proposition 3.26 we have

$$\langle\Phi, a\Psi\rangle = \langle\Phi a, \Psi\rangle = \int_{I(A^{**})} \Phi_1(av^*,v)d\mu(v)$$

$$= \int_{I(A^{**})} \Phi_1(v^*,va)d\mu(v) = \langle a\Phi,\Psi\rangle = \langle\Phi,\Psi a\rangle.$$

Hence $a\Psi = \Psi a$.

If $F \in A^*$, let

$$\Theta_F = {}^t\bar{\omega}(F) \in (A \hat{\otimes} A)^* = \mathcal{L}_2(A),$$

i.e., for all $x,y \in A$, $\Theta_F(x,y) = F(xy)$. Let F_1 be the uniquely determined extension of $F \in A^* = A^{***}$ to a normal linear functional on the von Neumann algebra A^{**}. If $v \in I(A^{**})$, v^*v is the unit of A^{**}. For $F \in A^*$ and $a \in A$,

$$<F, {}^{tt}_{\bar{\omega}}(\Psi)a> = <{}^{t}_{\bar{\omega}}(aF),\Psi> = <\Theta_{aF},\Psi>$$

$$= \int_{I(A^{**})} aF_1(v^*v)d\mu(v)$$

$$= \int_{I(A^{**})} <\underset{\sim}{1},aF_1>d\mu(v) = <a,F>.$$

Hence ${}^{tt}_{\bar{\omega}}(\Psi)a = a$.

Proposition 1.7 insures amenability of the algebra A. □

THEOREM 3.28: Every amenable von Neumann algebra is injective.

PROOF: (1) Let A be a von Neumann algebra operating on a Hilbert space H and suppose that the algebra is semifinite for a normal, faithful, semifinite trace θ on A_+. We assume θ to be extended to a linear functional on A and say that $T \in \mathcal{L}(H)$ has θ-finite rank if the central support of T is majorized by a projection p in A for which $\theta(p) < \infty$. The set F of all θ-finite rank operators in $\mathcal{L}(H)$ constitutes a closed A-submodule in $\mathcal{L}(H)$.

If $x \in A$, let $\|x\|_2 = \theta(x^*x)^{\frac{1}{2}}$. We consider the vector space Y formed by all linear functionals ϕ on F for which there exist $k = k_\phi > 0$ such that $|\phi(aTb)| \leq k \|a\|_2 \|T\| \|b\|_2$ whenever $a,b \in A \cap F$ and $T \in F$. We determine a norm $\|\cdot\|_Y$ on Y putting, for $\phi \in Y$,

$$\|Y\|_Y = \inf \{k \in R^*_+ : (\forall a \in A \cap F)(\forall b \in A \cap F)(\forall T \in F)$$

$$|\phi(aTb)| \leq k \|a\|_2 \|T\| \|b\|_2\}.$$

The unit ball of Y is compact, hence closed, in the $\sigma(Y,F)$-topology; Y is a Banach space and the topological dual space of a Banach space Y_* in which F is dense.

For all $a \in A \cap F$ and $y \in A$, we have $y^*y \leq \|y\|^2$; thus $a^*y^*ya \leq \|y\|^2 a^*a$ and

$$\|ya\|_2^2 = \theta(a^*y^*ya) \leqq \|y\|^2\,\theta(a^*a) = \|y\|^2\,\|a\|_2^2,$$

$$\|ya\|_2 \leqq \|y\|\,\|a\|_2.$$

Similarly, for all $b \in A \cap F$ and $y \in A$, we have $\|bx\|_2 \leqq \|b\|_2\,\|y\|$. Therefore, if $x,y \in A$ and $\phi \in Y$,

$$\|x\phi y\|_Y = \sup\{|\langle aTb, x\phi y\rangle| : (\forall a \in A \cap F)(\forall b \in A \cap F)(\forall T \in F)$$

$$\|a\|_2 \leqq 1,\ \|b\|_2 \leqq 1,\ \|T\| \leqq 1\}$$

$$= \sup\{|\langle yaTbx, \phi\rangle| : (\forall a \in A \cap F)(\forall b \in A \cap F)(\forall T \in F)$$

$$\|a\|_2 \leqq 1,\ \|b\|_2 \leqq 1,\ \|T\| \leqq 1\}$$

$$\leqq \|x\|\,\|y\|\,\|\phi\|_Y.$$

We conclude that Y is a Banach A-module.

As F is a Banach A-module, for every $x \in A$, the mapping $\underset{Y \to Y}{\phi \mapsto x\phi}$ is $\sigma(Y,F)$-continuous. As F is dense in Y_*, Y then constitutes a dual Banach A-module. Fix $T \in F$. If p is a projection in A majorizing the central support of T and such that $\theta(p) < \infty$, then, for all $x \in A$ and $\phi \in Y$,

$$|\phi(xT)| = |\phi(xpTp)| \leqq \|\phi\|_Y\,\|xp\|_2\,\|T\|\,\|p\|_2.$$

As the trace θ is normal, we conclude that the linear functional $x \mapsto \phi(xT)$ is normal on A. Hence the dual Banach A-module Y is normal, i.e., for all $T \in F$ and $\phi \in Y$, the linear functionals

$$x \mapsto \langle T, x\phi\rangle$$

$$x \mapsto \langle T, \phi x\rangle$$

are normal on A. The set $X = \{\phi \in Y : (\forall x \in A \cap F)\ \phi(x) = 0\}$ is $\sigma(Y,F)$-closed in Y; it constitutes a normal dual Banach A-submodule of Y.

As θ is normal, we may determine a family $\{\xi_i : i \in I\}$ of unitary vectors in H such that, for every $x \in A \cap F$,

$$\theta(x) = \sum_{i\in I} (x\xi_i|\xi_i) \quad \text{with} \quad \sum_{i\in I} |(x\xi_i|\xi_i)| < \infty .$$

If $T = T_1 + iT_2$ for real operators T_1, T_2 in F, one may determine a projection q in A such that $\theta(q) < \infty$ and $-\|T_j\| q \leqq T_j \leqq \|T_j\| q$ for $j = 1,2$. Hence $\sum_{i\in I} |(T\xi_i|\xi_i)|$ exists and one may determine a linear functional θ_0 on F putting

$$\theta_0(T) = \sum_{i\in I} (T\xi_i|\xi_i),$$

$T \in F$. Thus

$$\theta_0(T) \geqq 0$$

for $T \in F_+$, and

$$\theta_0(T) = \theta(T)$$

for $T \in A \cap F$. As F is a *-subalgebra of $\mathcal{L}(H)$,

$$|\theta_0(U^*V)| \leqq \theta_0(U^*U)\ \theta_0(V^*V)$$

whenever $U,V \in F$ ([D'] 2.1.2).

Let $T \in F$ and $a,b \in A \cap F$. We have

$$\theta_0(aTT^*a^*) = \sum_{i\in I} \|T^*a^*\xi_i\|^2$$

$$\leqq \|T\|^2 \sum_{i\in I} \|a^*\xi_i\|^2 = \|T\|^2\ \theta_0(aa^*)$$

and

$$|\theta_0(aTb)|^2 \leqq \theta_0(aTT^*a^*)\theta_0(b^*b)$$

$$\leqq \|T\|^2\ \theta_0(aa^*)\theta_0(b^*b).$$

Then we obtain

$$|\theta_0(aTb)| \leq \|T\| \; \|a\|_2 \; \|b\|_2.$$

Therefore $\theta_0 \in Y$ and $\|\theta_0\|_Y \leqq 1$.

We now consider the continuous derivation D of A into Y defined by

$$D(x) = x\theta_0 - \theta_0 x,$$

$x \in A$. As $A \cap F$ is an A-module, for all $x \in A$ and $y \in A \cap F$,

$$\theta_0(xy-yx) = \theta(xy-yx) = 0.$$

Hence D is a derivation into X; it is also normal. As A is amenable, there exists $\phi \in X$ such that $D = d_1\phi$. Let $\psi = \theta_0 - \phi \in Y$. We have

$$\theta(x) = \psi(x) \tag{13}$$

whenever $x \in A \cap F$. If $x \in A$,

$$x\theta_0 - \theta_0 x = D(x) = d_1\phi(x) = x\phi - \phi x,$$

hence

$$x\psi = \psi x. \tag{14}$$

Let p be a nonzero projection in A such that $\theta(p) < \infty$ and let $H_1 = pH$; the von Neumann algebra A_p operates on H_1. As the trace θ is faithful, so is its restriction θ_1 to A_p. As $\mathcal{L}(H_1) \simeq \{T \in \mathcal{L}(H) : pT = T = Tp\}$, we deduce from (13) that the restriction ψ_1 of ψ to $\mathcal{L}(H_1)$ is nonzero. By (14), for every $x \in A_p$,

$$x\psi_1 = \psi_1 x. \tag{15}$$

From (13) we obtain

$$\theta_1(x) = \psi_1(x)$$

for every $x \in A_p \subset A \cap F$. In the relations (15) and (16) we may assume ψ_1 to be hermitian. We conclude that, for every $x \in A_p$ and every $T \in \mathcal{L}(H_1)$,

$$(\psi_1)_+ (xT) = (\psi_1)_+ (Tx),$$

and, for every $x \in (A_p)_+$,

$$(\psi_1)_+(x) \geqq \theta_1(x).$$

Let $\rho = \|(\psi_1)_+\|^{-1}(\psi_1)_+ \in \mathcal{L}(H_1)^*$. As $\mathcal{L}(H_1)$ is a von Neumann algebra, ρ is a weak-*-limit of normal states on $\mathcal{L}(H_1)$. Let $a_1,\dots,a_n$, $b_1,\dots,b_n \in A_p$. Given $\varepsilon > 0$, there exists a normal state ρ' on $\mathcal{L}(H_1)$ such that

$$\left| \left\| \sum_{j=1}^{n} b_j^{*}\rho' a_j \right\| - \left|\left\langle \sum_{j=1}^{n} a_j\, b_j^{*}, \rho\right\rangle\right| \right|$$

$$= \left| \left\langle \underset{\sim}{1}, \sum_{j=1}^{n} b_j^{*}\rho' a_j\right\rangle\right| - \left|\left\langle \sum_{j=1}^{n} a_j b_j^{*}, \rho\right\rangle\right| \Bigg|$$

$$= \left|\left|\left\langle \sum_{j=1}^{n} a_j b_j^{*}, \rho'\right\rangle\right| - \left|\left\langle \sum_{j=1}^{n} a_j b_j^{*}, \rho\right\rangle\right|\right|$$

$$\leqq \left|\left\langle \sum_{j=1}^{n} a_j b_j^{*}, \rho'-\rho\right\rangle\right| < \varepsilon$$

and, for every $j = 1,\dots,n$,

$$\|\rho' b_j - b_j\rho'\|$$

$$\leqq \|(\rho'-\rho)b_j - b_j(\rho'-\rho)\| + \|\rho\, b_j - b_j\rho\|$$

$$\leqq |\langle b_j, \rho'-\rho\rangle| + |\langle b_j, \rho'-\rho\rangle| < \varepsilon .$$

So there exists $T \in n(H_1)$ such that $\|T\|_{HS} = 1$, $\|b_j T - T b_j\|_{HS} < \varepsilon$ whenever $j = 1,\dots,n$, and $\left\|\sum_{j=1}^{n} a_j T b_j^{*}\right\|_{HS} > \left|\left\langle \sum_{j=1}^{n} a_j b_j^{*}, \rho\right\rangle\right| - \varepsilon$. Making use of the identification of $n(H_1)$ with $H_1 \otimes H_1^c$, we conclude that

$$|\langle \sum_{j=1}^{n} a_j b_j^*, \rho\rangle| \leq \|\sum_{j=1}^{n} a_j \otimes b_j^c\|.$$

Therefore the mapping $\bar{\omega}$ defined on $A_p \otimes \widetilde{A_p}$ is continuous with respect to the topology defined by $\|\cdot\|_{min}$. Theorem 3.18 implies semidiscreteness of the von Neumann algebra A_p. By Proposition 3.19 and Corollary 3.23, A_p is injective in $\mathcal{L}(H_1)$. Then the algebra A itself is injective in $\mathcal{L}(H)$.

2) Let A be a von Neumann algebra of type III; A is the crossed product of a von Neumann algebra A_1 of type II_∞, hence semifinite, by a one-parameter group. The algebra A_1 is generated by a von Neumann algebra B, that is isomorphic to A, and a one-parameter group $\{U_t : t \in \mathbb{R}\}$ of unitary representation of $\mathbb{R}$ such that $U_t B U_t^* = B$ whenever $t \in \mathbb{R}$ ([67] p. 60). We show amenability of A_1; then A_1 is injective by 1). Thus so also is A.

Let X be a Banach A_1-module, hence a Banach B-module, and consider a continuous derivation D_0 of A_1 into X*. As B is amenable, there exists $b \in B$ such that $D_0(x) = d_1 b(x)$ whenever $x \in B$. For the continuous derivation $D = D_0 - d_1 b$ of A_1 into X* we have $D(x) = 0$ whenever $x \in B$.

For $s \in \mathbb{R}$, let

$$\xi_s = U_s D(U_s^{-1}) \in X^*.$$

If $y \in B$, $U_s^{-1} y U_s \in B$, hence $D(U_s^{-1} y U_s) = 0$ and

$$\begin{aligned} yU_s D(U_s^{-1}) &= (U_s U_{s^{-1}}) y U_s D(U_s^{-1}) = U_s (U_s^{-1} y U_s) D(U_s^{-1}) \\ &= U_s (D(U_s^{-1} y U_s) U_s^{-1} + (U_s^{-1} y U_s) D(U_s^{-1})) \\ &= U_s D(U_s^{-1} y) \\ &= U_s (D(U_s^{-1}) y + U_s^{-1} D(y)) = U_s D(U_s^{-1}) y, \end{aligned}$$

i.e.,

$$y\xi_s = \xi_s y. \tag{17}$$

For all $s,t \in \underset{\sim}{R}$,

$$U_t\xi_s = U_tU_sD(U_s^{-1}) = U_{t+s}D(U_{s+t}^{-1}U_t)$$

$$= U_{t+s}D(U_{s+t}^{-1})U_t + D(U_t);$$

$$U_t\xi_s = \xi_{s+t}\, U_t + D(U_t). \tag{18}$$

As the abelian group $\tilde{R}$ is amenable in the discrete topology, there exists an increasing sequence (F_n) of finite subsets in $\underset{\sim}{R}$ such that $(\frac{1}{\text{card } F_n} \sum_{s \in F_n} \xi_s)$ converges to $\xi \in X^*$ in the weak-*-topology ([P] Theorem 24.13).

Then, for every $y \in B$, by (17),

$$D(y) = 0 = d_1\xi(y)$$

and, for every $t \in \underset{\sim}{R}$, by (18),

$$U_t\xi = \xi U_t + D(U_t),$$

i.e.,

$$D(U_t) = d_1\xi(U_t).$$

Thus the derivation D is inner and so is D_0. We conclude that A_1 is amenable. □

<u>LEMMA 3.29</u>: Let A be a finite injective von Neumann algebra with center Z and let Φ be a central trace defined on A. Given $a_1,\dots,a_n$, $b_1,\dots,b_n \in A$ and $\varepsilon > 0$, there exists a normal state f on Z such that

$$|f(\Phi(\sum_{i=1}^{n} a_ib_i^*))| \leq \|\sum_{i=1}^{n} a_i \otimes b_i^c\|_{\min} + \varepsilon.$$

<u>PROOF</u>: Let H be the Hilbert space on which A operates and let p be the projection of norm 1 mapping $\mathcal{L}(H)$ onto A. Consider the product $(\mathcal{L}(H)_*)^m$

formed by m copies of $\mathcal{L}(H)_*$ and equipped with the norm defined by

$$\|(\phi_1,\dots,\phi_m)\| = \sum_{i=1}^{m} \|\phi_i\| \text{ for } (\phi_1,\dots,\phi_m) \in (\mathcal{L}(H)_*)^m.$$

Let $v_1,\dots,v_m \in A_u$ and denote by C the set of all the elements $(\psi - v_1^*\psi v_1,\dots,\psi - v_m^*\psi v_m)$ where ψ runs over the set of all normal states on $\mathcal{L}(H)$; C is convex in $(\mathcal{L}(H)_*)^m$. Let g be a normal state on Z; then $\Psi = g\circ\Phi\circ p$ is a state on $\mathcal{L}(H)$. For every $v \in A_u$ and every $x \in \mathcal{L}(H)$, $p(vxv^*) = vp(x)v^*$ ([T] Chapter III, Theorem 3.4), hence $\Phi\circ p(vxv^*) = \Phi\circ p(x)$ and $\Psi(vxv^*) = \Psi(x)$. Therefore 0 belongs to the closed convex hull of C; for every $\varepsilon > 0$, there exists a normal state Ψ_0 on $\mathcal{L}(H)$ such that $\|v_j\Psi_0 - \Psi_0 v_j\| = \|\Psi_0 - v_j^*\Psi_0 v_j\| < \varepsilon$ whenever $j = 1,\dots,m$. There exists a tracial operator $\rho \in \mathcal{L}(H)_+$ such that $\Psi_0 = \mathrm{Tr}\,\rho$([S] 1.15.3). Thus, for all $x \in A$ and $j = 1,\dots,m$,

$$v_j\Psi_0 v_j^*(x) = \mathrm{Tr}\,\rho\,(v_j^* x v_j) = \mathrm{Tr}\,v_j\rho v_j^*(x)$$

and

$$|\mathrm{Tr}(v_j\rho v_j^* - \rho)(x)| \leqq \varepsilon\ \|x\|.$$

So, for every $\varepsilon > 0$, one may deduce the existence of a nonzero finite-rank projection p_o in $\mathcal{L}(H)$ such that

$$\|v_j\, p_o - p_o v_j\|_{HS} \leqq \varepsilon\,\|p_o\|_{HS}$$

whenever $j = 1,\dots,m$. Then also if F is any finite subset in A and $\varepsilon > 0$, there exists a nonzero finite-rank projection p in $\mathcal{L}(H)$ such that

$$\|xp - px\|_{HS} \leqq \varepsilon\ \|x\|_{HS}$$

whenever $x \in F$.

Let $a_1,\dots,a_n$, $b_1,\dots,b_n \in A$ such that $\|a_i\| \leqq 1$ whenever $i = 1,\dots,n$ and $\varepsilon > 0$. We claim the existence of $w_1,\dots,w_s \in A_u$ and $\delta \in]0,1[$ such that, for $c = \sum_{i=1}^{n} a_i b_i^*$ and a state ϕ on A satisfying

$$\|w_k\phi - \phi w_k\| < \delta$$

whenever $k = 1,\ldots,s$, the relation

$$|\phi(c - \Phi(c))| \leqq \varepsilon/(n+1)$$

holds.

Otherwise there exists $\alpha > 0$ such that, to any finite subset F in A_u and every $m \in \underset{\sim}{N}^*$, one may associate a state $\phi_{F,m}$ on A satisfying $\|v\phi_{F,m} - \phi_{F,m} v\| < 1/m$ whenever $v \in F$, and $|\phi_{F,m}(c - \Phi(c))| > \alpha$. But a weak-*-limit state ϕ of the net $(\phi_{F,m})$ is a trace; then $\phi = \phi\circ\Phi$ ([D] Chapter III, §5, Proposition 3) and a contradiction arises.

We may determine a nonzero finite-rank projection $q \in \mathcal{L}(H)_+$ such that

$$\|b_i q - q b_i\|_{HS} \leqq \varepsilon/(n+1) \|q\|_{HS}$$

whenever $i = 1,\ldots,n$, and

$$\|w_k q - q w_k\|_{HS} \leqq (\delta/3) \|q\|_{HS}$$

whenever $k = 1,\ldots,s$. Let

$$\phi(x) = \langle xq,q\rangle_{HS} / \langle q,q\rangle_{HS},$$

$x \in A$.

For $k = 1,\ldots,s$, put

$$q_k = w_k^* q w_k;$$

hence $w_k^* q = q_k w_k^*$ and

$$\|q_k - q\|_{HS} = \|w_k^*(qw_k - w_k q)\|_{HS} \leqq \|qw_k - w_k q\|_{HS}$$
$$\leqq (\delta/3) \|q\|_{HS}.$$

For every $x \in A$,

$$\langle w_k x w_k^* q, q\rangle_{HS} = \langle x w_k^* q, w_k^* q\rangle_{HS}$$

$$= \langle x q_k w_k^*, q_k w_k^*\rangle_{HS} = \langle x q_k, q_k\rangle_{HS};$$

hence

$$|\langle w_k x w_k^* q, q\rangle_{HS} - \langle xq, q\rangle_{HS}|$$

$$= |\langle x q_k, q_k\rangle_{HS} - \langle xq, q\rangle_{HS}|$$

$$\leqq |\langle x q_k, q_k - q\rangle_{HS}| + |\langle x q_k - xq, q\rangle_{HS}|$$

$$\leqq 2\,\|x\|\,\|q_k - q\|_{HS}\,\|q\|_{HS} \leqq 2(\delta/3)\,\|x\|\;\|q\|_{HS}^2$$

and

$$|\phi(w_k x w_k^* - x)| \leqq \delta\,\|x\|.$$

Therefore

$$\|w_k\phi - \phi w_k\| \leqq \delta$$

and we obtain

$$|\phi(\Phi(c) - c)| \leqq \varepsilon/(n+1) = \varepsilon'.$$

We have also

$$\Big|\sum_{i=1}^{n} \langle a_i b_i^* q - a_i q b_i^*, q\rangle_{HS}\Big|$$

$$\leqq \sum_{i=1}^{n} \|a_i\|\;\|b_i^* q - q b_i^*\|_{HS}\,\|q\|_{HS}$$

$$\leqq \sum_{i=1}^{n} \|q b_i - b_i q\|_{HS}\;\|q\|_{HS}$$

$$\leqq n\varepsilon'\;\|q\|_{HS},$$

hence

$$\left|\frac{1}{\langle q,q\rangle_{HS}} \sum_{i=1}^{n} \langle a_i q b_i^*, q\rangle_{HS} - \phi(c)\right| \leq n\varepsilon'$$

and

$$|\phi(\Phi(c))| \leq |\phi(\Phi(c)-c)| + \left|\phi(c) - \frac{1}{\langle q,q\rangle_{HS}} \sum_{i=1}^{n} \langle a_i q b_i^*, q\rangle_{HS}\right|$$

$$+ \frac{1}{\langle q,q\rangle_{HS}} \left|\sum_{i=1}^{n} \langle a_i q b_i^*, q\rangle_{HS}\right| \tag{19}$$

$$\leq \varepsilon + \frac{1}{\langle q,q\rangle_{HS}} \left|\sum_{i=1}^{n} \langle a_i q b_i^*, q\rangle_{HS}\right|.$$

Let $B = qA$, $H_1 = qH$, $d = \dim H_1$ and consider an othonormal basis $(\eta_1,\ldots,\eta_d)$ of H_1. Put

$$\eta = d^{-1/2} \sum_{t=1}^{d} \eta_t \otimes \eta_t^c \in H_1 \otimes H_1^c;$$

$\|\eta\| = 1$ and, for every $x \in B$, $((x \otimes \underset{\sim}{1})\eta|\eta) = \phi(x)$. Let also $a_i' = qa_iq$, $b_i' = qb_iq$ for $i = 1,\ldots,n$. We consider the isometric involution on $H_1 \otimes H_2^c$ defined by $J_0(\xi_1 \otimes \xi_2^c) = \xi_2 \otimes \xi_1^c$ for $\xi_1 \in H_1$, $\xi_2^c \in H_2^c$. We have $J_0\eta = \eta$ and, in the finite-dimensional Hilbert space H_1, for $x \in B$, $J_0(x \otimes \underset{\sim}{1})J_0\eta = (x^* \otimes \underset{\sim}{1})\eta$; therefore

$$\phi\left(\sum_{i=1}^{n} a_i' b_i'^*\right) = \left(\left(\sum_{i=1}^{n} a_i' b_i'^* \otimes \underset{\sim}{1}\right)\eta|\eta\right)$$

$$= \sum_{i=1}^{n} ((a_i' \otimes \underset{\sim}{1})J_0(b_i' \otimes \underset{\sim}{1})J_0\eta|\eta) = \sum_{i=1}^{n} (a_i' \otimes b_i'^c\eta|\eta).$$

As $H_1 \otimes H_1^c \subset H \otimes H^c$, we have also

$$(q \otimes q^c)\left(\sum_{i=1}^{n} a_i \otimes b_i^c\right)(q \otimes q^c) = \sum_{i=1}^{n} a_i' \otimes b_i'^c;$$

thus

$$\frac{1}{\langle q,q\rangle_{HS}} \left|\sum_{i=1}^{n} \langle a_i q b_i^*, q\rangle\right|_{HS} = \frac{1}{\langle q,q\rangle_{HS}} \left|\sum_{i=1}^{n} \langle q a_i q q b_i^* q, q\rangle_{HS}\right| \tag{20}$$

$$= \left|\phi\left(\sum_{i=1}^{n} a_i^! b_i^{!*}\right)\right| = \left|\sum_{i=1}^{n} (a_i \otimes b_i^c \eta|\eta)\right| \leq \left\|\sum_{i=1}^{n} a_i \otimes b_i^c\right\|_{min}.$$

From (19) and (20) we deduce that

$$|\phi(\Phi(c))| \leq \varepsilon + \left\|\sum_{i=1}^{n} a_i \otimes b_i^c\right\|_{min}.$$

It suffices now to choose for f the restriction of ϕ to Z. □

LEMMA 3.30: Let A be a finite injective von Neumann algebra and let T be a central trace on A. Then, for all $a_1,\ldots,a_n$, $b_1,\ldots,b_n \in A$,

$$\left\|T\left(\sum_{i=1}^{n} a_i b_i^*\right)\right\| \leq \left\|\sum_{i=1}^{n} a_i \otimes b_i^c\right\|_{min}.$$

PROOF: Let $\alpha = \left\|\sum_{i=1}^{n} a_i \otimes b_i^c\right\|_{min}$. Suppose that $\left\|T\left(\sum_{i=1}^{n} a_i b_i^*\right)\right\| > \alpha$. There exist a nonzero spectral projection p in the center of A and $\beta \in \mathbb{C}$ such that $|\beta| > \alpha$ and

$$\left\|pT\left(\sum_{i=1}^{n} a_i b_i^*\right) - \beta p\right\| < \frac{1}{2}(|\beta| - \alpha).$$

Let $\varepsilon = \frac{1}{2}(|\beta| - \alpha)$. As the von Neumann algebra A is injective, so is A_p. By Lemma 3.29 one may determine a normal state f on the center of A_p such that

$$\left|f\left(pT\left(\sum_{i=1}^{n} a_i b_i^*\right)\right)\right| \leq \left\|\sum_{i=1}^{n} p a_i p \otimes (p b_i p)^c\right\|_{min} + \varepsilon$$

$$= \left\|(p \otimes p^c)\left(\sum_{i=1}^{n} a_i \otimes b_i^c\right)(p \otimes p^c)\right\|_{min} + \varepsilon$$

$$\leq \left\|\sum_{i=1}^{n} a_i \otimes b_i^c\right\|_{min} + \varepsilon = \alpha + \frac{1}{2}(|\beta| - \alpha) = \frac{1}{2}(|\beta| + \alpha).$$

But

$$|f(pT(\sum_{i=1}^{n} a_i b_i^*)) - \beta| = |f(pT(\sum_{i=1}^{n} a_i b_i^*) - \beta p)|$$

$$\leq \|pT(\sum_{i=1}^{n} a_i b_i^*) - \beta p\| < \varepsilon;$$

hence

$$|f(pT(\sum_{i=1}^{n} a_i b_i^*))| > |\beta| - \varepsilon = \frac{1}{2}(|\beta| + \alpha).$$

We come to a contradiction. □

PROPOSITION 3.31: Every finite injective von Neumann algebra A is semi-discrete.

PROOF: With respect to Proposition 3.10 we may assume A to be indecomposable and consider a normal faithful state θ on the center of A. If T is a normal central trace on A, the normal faithful state $f = \theta \circ T$ reduces to a trace on A_+. By Lemma 3.30, for all $a_1,\ldots,a_n$, $b_1,\ldots,b_n \in A$,

$$|f(\sum_{i=1}^{n} a_i b_i^*)| \leq \|T(\sum_{i=1}^{n} a_i b_i^*)\| \leq \|\sum_{i=1}^{n} a_i \otimes b_i^c\|_{min}.$$

We consider the representation associated to f, on a standard von Neumann algebra, for a cyclic unitary vector ξ of a Hilbert space H;

$$f(x) = (x\xi|\xi),$$

$x \in A$. Then

$$|\sum_{i=1}^{n} (a_i Jb_i J\xi|\xi)| \leq \|\sum_{i=1}^{n} a_i \otimes Jb_i J\|_{min};$$

for all $x_1,\ldots,x_n \in A$, $y_1,\ldots,y_n \in \tilde{A}$,

$$|(\sum_{i=1}^{n} x_i y_i \xi|\xi)| \leq \|\sum_{i=1}^{n} x_i \otimes y_i\|_{min}.$$

thus the linear functional

$$w : x \to (\bar{\omega}(x)\xi|\xi),$$

defined on the algebraic tensor product of A and $A^{\sim}$, is continuous for the norm $\|\cdot\|_{min}$; it admits a unique extension to a state on $A \otimes_{min} A^{\sim}$.

Let $x,y \in A \otimes A^{\sim}$. Then

$$w(y^*x^*xy) \leq \|x^*x\|_{min}\, w(y^*y) = \|x\|^2_{min}\, w(y^*y)$$

([D'] 2.1.5 (ii)); moreover, as $\bar{\omega}$ is defined on $A \otimes A^{\sim}$, we have $\bar{\omega}(y^*y) = \bar{\omega}(y^*)\bar{\omega}(y)$ and

$$w(y^*y) = (\bar{\omega}(y^*y)\xi|\xi) = (\bar{\omega}(y^*)\bar{\omega}(y)\xi|\xi)$$

$$= \|\bar{\omega}(y)\xi\|^2,$$

$$\|\bar{\omega}(x)\bar{\omega}(y)\xi\|^2 = w(y^*x^*xy) \leq \|x\|^2_{min}\, \|\bar{\omega}(y)\xi\|^2.$$

Hence, for every $x \in A \otimes A^{\sim}$,

$$\|\bar{\omega}(x)\| \leq \|x\|_{min}$$

as $\quad \{\bar{\omega}(y)\xi : y \in A \otimes A^{\sim}\}^- = \mathcal{H}.$

The statement now follows from Theorem 3.18. □

<u>COROLLARY 3.32</u>: Every semifinite injective von Neumann algebra is semidiscrete.

<u>PROOF</u>: Any semifinite injective von Neumann algebra admits a decomposition into a direct sum formed by a von Neumann algebra of type I, a von Neumann algebra of type II_1 and a von Neumann algebra of type II_∞. Any type I von Neumann algebra is semidiscrete and by Proposition 3.31 the component of type II_1 is also semidiscrete. Thus, by Proposition 3.10 it suffices to show that any injective von Neumann algebra A of type II_∞ is semidiscrete.

1) Let p be a nonzero finite projection of A. By Zorn's lemma we may determine a maximal family $\{p_i : i \in I\}$ of pairwise orthogonal projections in A that are all equivalent to p. As A is of type II_∞, the family is infinite. Let $q = \underset{\sim}{1} - \sum_{i \in I} p_i$; we may write

$$p \lesssim \sum_{i \in I} p_i = \underset{\sim}{1} - q. \tag{21}$$

There exists a central projection z such that $qz \lesssim pz$ and $(\underset{\sim}{1}-z)p \lesssim (\underset{\sim}{1}-z)q$ ([D] Chapter III, §1, Theorem 1). If we had $z = 0$, then

$$p \lesssim q. \tag{22}$$

From (21) and (22) we would deduce that $p = p(\underset{\sim}{1}-q) + pq = p + p$; a contradiction would arise. Therefore $z \neq 0$.

We have

$$\sum_{i \in I} zp_i \lesssim zq + \sum_{i \in I} zp_i = z(q + \sum_{i \in I} p_i) = z.$$

On the other hand, as $qz \lesssim pz$,

$$z = qz + (\sum_{i \in I} p_i)z \lesssim pz + \sum_{i \in I} p_i z = \sum_{i \in I} p_i z.$$

Thus we have $z \sim \sum_{i \in I} zp_i$ ([D] Chapter III, §1, Proposition 1). Proceeding to an exchange of equivalent projections, if necessary, we may assume that

$$z = \sum_{i \in I} zp_i.$$

Let $H = \ell^2(I)$. Then A_z is spatially isomorphic to $\mathcal{L}(H) \bar{\otimes} A_{zp}$ ([D] Chapter I, §2, Proposition 5); A_{zp} is injective and finite. Thus by Proposition 3.31, A_{zp} is semidiscrete. We deduce from Proposition 3.12 that A_z is semidiscrete.

2) By Zorn's lemma there exists a maximal family $\{z_s : s \in S\}$ of nonzero central projections that are pairwise orthogonal and such that A_{z_s} is semidiscrete for every $s \in S$. If $q = \sum_{s \in S} z_s$ differs from id_A, then, for the

central projection 1-q, the von Neumann algebra A_{1-q} is nontrivial and injective; hence, by 1), there exists a nonzero central projection z such that $z \leq 1-q$ and A_z is semidiscrete. We contradict the maximality of the family $\{z_s : s \in S\}$. Thus $\sum_{s \in S} z_s = id_A$ and the von Neumann algebra $A = \underset{s \in S}{\oplus} A_{z_s}$ is semidiscrete by Proposition 3.10. $\square$

The group $\underset{\sim}{R}$ coincides with its dual group ([41] 23.27(e)).

LEMMA 3.33: Every von Neumann algebra A admits a projection of norm 1 mapping $A \underset{\alpha}{\otimes} \underset{\sim}{R}$ onto $\{\lambda_x : x \in A\}$.

PROOF: By continuity of $\hat{\alpha}$, for all $n \in N*$ and $x \in A$, one may consider $\int_{-n}^{n} \hat{\alpha}_t(x)dt$. If $n \in N*$ and $x \in A$, we put

$$\Psi_n(x) = \frac{1}{2n} \int_{-n}^{n} \hat{\alpha}_t(x)dt;$$

$\|\Psi_n(x)\| \leq \|x\|$ and, for every $s \in \underset{\sim}{R}$, $\Psi_n \circ \hat{\alpha}_s = \hat{\alpha}_s \circ \Psi_n$. Then

$$\Psi_n(\hat{\alpha}_s(x)) - \Psi_n(x) = \frac{1}{2n} \int_{-n}^{n} (\hat{\alpha}_{t+s}(x) - \hat{\alpha}_t(x))dt$$

$$= \frac{1}{2n} (\int_{-n+s}^{n+s} - \int_{-n}^{n} \hat{\alpha}_t(x)dt)$$

$$= \frac{1}{2n} (\int_{n}^{n+s} + \int_{-n}^{-n+s} \hat{\alpha}_t(x)dt)$$

and

$$\|\Psi_n(\hat{\alpha}_s(x)) - \Psi_n(x)\| \leq \frac{1}{n} |s|.$$

Thus

$$\lim_{n \to \infty} \|\Psi_n(\hat{\alpha}_s(x)) - \Psi_n(x)\| = 0.$$

Let Ψ be a weak-*-limit of (Ψ_n). For all $x \in A$ and $s \in \underset{\sim}{R}$, $\|\Psi(x)\| \leq \|x\|$

and $\Psi(x) = \Psi(\hat{\alpha}_s(x)) = \hat{\alpha}_s \Psi(x)$. As $\{\lambda_x : x \in A\}$ is the subalgebra of $A \otimes_\alpha \underset{\sim}{R}$ consisting of the points that are invariant under the action of $\{\hat{\alpha}_s : s \in \underset{\sim}{R}\}$, we conclude that Ψ is the desired projection of norm 1. □

<u>THEOREM 3.34</u>: Every injective von Neumann algebra A is semidiscrete.

<u>PROOF</u>: The algebra A admits a decomposition $A_1 \oplus A_2$ for a semifinite injective von Neumann algebra A_1 and an injective von Neumann algebra A_2 of type III. Taking into account Corollary 3.32 and Proposition 3.10 we may suppose A to be of type III.

Let H be the Hilbert space on which A operates and consider the modular action α of $\underset{\sim}{R}$. As $(A \otimes_\alpha \underset{\sim}{R}) \otimes_{\hat{\alpha}} \underset{\sim}{R}$ is isomorphic to $A \bar{\otimes} \mathcal{L}(L^2(\underset{\sim}{R}))$ and the von Neumann algebra A is of type III, the crossed product is isomorphic to A; $A \otimes_\alpha \underset{\sim}{R}$ is a semifinite von Neumann algebra. By Lemma 3.33 there exists a projection p of norm 1 mapping the crossed product onto the algebra $\{\lambda_x : x \in A \otimes_{\tilde{\alpha}} \underset{\sim}{R}\}$. Injectivity of A implies the existence of a projection p', with norm 1, mapping $\mathcal{L}(H \otimes L^2(\underset{\sim}{R}) \otimes L^2(\underset{\sim}{R}))$ onto the crossed product $(A \otimes_\alpha \underset{\sim}{R}) \otimes_{\tilde{\alpha}} \underset{\sim}{R}$. Thus $p \circ p'$ is also a projection of norm 1; the algebra $A \otimes_\alpha \underset{\sim}{R}$ is injective. By Corollary 3.32, the semifinite von Neumann algebra $A \otimes_\alpha \underset{\sim}{R}$ is then semidiscrete. By Theorem 3.18, if $a_1, \ldots, a_n \in A \otimes_\alpha \underset{\sim}{R}$ and $b_1, \ldots, b_n \in (A \otimes_\alpha \underset{\sim}{R})^{\sim}$,

$$\left\| \sum_{i=1}^{n} a_i b_i \right\| \leq \left\| \sum_{i=1}^{n} a_i \otimes b_i \right\|_{min}.$$

If $y \in A^{\sim}$, put $\mu_y = y \otimes \underset{\sim}{1}$. Thus, for every $f \in L^2(\underset{\sim}{R}, H) \simeq H \otimes L^2(\underset{\sim}{R})$ and every $s \in \underset{\sim}{R}$,

$$\mu_y f(s) = y f(s);$$

$$\{\mu_y : y \in A^{\sim}\} \subset (A \otimes_\alpha \underset{\sim}{R})^{\sim}.$$

We consider $A \otimes A^{\sim}$. For $x_1, \ldots, x_n \in A$, $y_1, \ldots, y_n \in A^{\sim}$,

$$\left\| \sum_{i=1}^{n} \lambda_{x_i} \mu_{y_i} \right\| \leq \left\| \sum_{i=1}^{n} \lambda_{x_i} \otimes \mu_{y_i} \right\|_{min}$$

and, as λ and μ are faithful representations,

$$\left\| \sum_{i=1}^{n} \lambda_{x_i} \mu_{y_i} \right\| \leq \left\| \sum_{i=1}^{n} x_i \otimes y_i \right\|_{\min}.$$

For every $\xi \in H$ and every H-valued continuous function f on $\underset{\sim}{R}$ with compact support,

$$\left\| \sum_{i=1}^{n} \lambda_{x_i} \mu_{y_i} (\xi \otimes f) \right\|^2 = \int_{\underset{\sim}{R}} \left\| \sum_{i=1}^{n} \alpha_{-t}(x_i) y_i \xi \right\|^2 |f(t)|^2 dt$$

$$\leq \left\| \sum_{i=1}^{n} \lambda_{x_i} \mu_{y_i} \right\|^2 \|\xi\|^2 \|f\|_2^2 \leq \left\| \sum_{i=1}^{n} x_i \otimes y_i \right\|_{\min}^2 \quad \|\xi\|^2 \quad \|f\|_2^2.$$

As f may be chosen arbitrarily and the mapping $t \mapsto \sum_{i=1}^{n} \alpha_{-t}(x_i) y_i \xi$ is continuous, we must have

$$\left\| \sum_{i=1}^{n} \alpha_{-t}(x_i) y_i \right\|^2 \leq \left\| \sum_{i=1}^{n} x_i \otimes y_i \right\|_{\min}^2$$

whenever $t \in \underset{\sim}{R}$. For $t = 0$, we obtain

$$\left\| \sum_{i=1}^{n} x_i y_i \right\| \leq \left\| \sum_{i=1}^{n} x_i \otimes y_i \right\|_{\min}.$$

Theorem 3.18 yields the statement. □

The final conclusion now is that, for a von Neumann algebra over a separable Hilbert space, amenability is equivalent to the other fundamental properties via the following diagram:

$$\begin{array}{ccc} \text{amenable} & \overset{3.28}{\Longrightarrow} & \text{injective} \\ 3.27 \Uparrow & & \Downarrow 3.34 \\ \text{nuclear} & \underset{3.19}{\Longleftarrow} & \text{semidiscrete} \end{array}$$

Comments

We consider some supplementary properties for von Neumann algebras.

DEFINITION 3.35: Let A be a von Neumann algebra over the Hilbert space H. We say that A is approximately finite-dimensional if there exists an increasing sequence of finite-dimensional *-subalgebras of A containing id_H and the union of which is strongly dense in A.

Every factor of type I is approximately finite-dimensional.

We formulate a sufficient condition for a properly infinite factor A to be approximately finite-dimensional. If $x_1,\ldots,x_n \in A$, ω is a normal state on A, and $\varepsilon > 0$, there exist a finite-dimensional subfactor F in A and $y_1,\ldots,y_n \in F$ such that $\|x_k - y_k\|_\omega^\# < \varepsilon$ whenever $k = 1,\ldots,n$ ([26] Theorem 3). One may assume $x_1,\ldots,x_n$ to be unitary elements.

DEFINITION 3.36: Let A be a von Neumann algebra over a Hilbert space H. The algebra A is said to have property (P) if, for every $T \in \mathcal{L}(H)$, $A^{\sim}$ admits a nonvoid intersection with the closed convex hull of $\{vTv^* : v \in A_u\}$.

PROPOSITION 3.37: Let A be a von Neumann algebra over the Hilbert space H and let $G = A_u$.

1) If A is approximately finite-dimensional, then G is amenable.

2) If G is amenable, A admits property (P).

PROOF: 1) Let (A_n) be an increasing sequence of unital finite-dimensional *-subalgebras in A such that $R = \bigcup_n A_n$ is strongly dense in A. As the set R_u of unitary elements in R is the union of an increasing sequence of compact, hence amenable, groups, $\overline{R_u}$ is an amenable group ([P] Proposition 13.6). But $\overline{R_u} = \bar{R}_u$ ([T] Chapter II, Theorem 4.11). Thus G is amenable.

2) Let M be a left invariant mean on C(G).

If $T \in \mathcal{L}(H)$ and $\xi, \eta \in H$, we consider

$$f = f_{T,\xi,\eta} : G \to \underset{\sim}{C}$$
$$v \mapsto (vTv^*\xi|\eta)$$

For all $v,w \in G$, we have

$$|f(v) - f(w)| \leq |(vTv^*(id_H - vw^*)\xi|\eta)| + |((id_H - wv^*)vTw^*\xi|\eta)|$$

$$= |(vTv^*(id_H - vw^*)\xi|\eta)| + |(vTw^*\xi|(id_H - vw^*)\eta)|$$

$$\leq \|vTv^*\| \ \|id_H - vw^*\| \ \|\xi\| \ \|\eta\| + \|vTw^*\| \ \|id_H - vw^*\| \ \|\xi\| \ \|\eta\|$$

$$\leq 2\|T\| \|id_H - vw^*\| \ \|\xi\| \ \|\eta\| .$$

Thus $f \in C(G)$. By Riesz's theorem, if $T \in \mathcal{L}(H)$, there exists $\Phi_T \in \mathcal{L}(H)$ such that

$$(\Phi_T\xi|\eta) = M(f_{T,\xi,\eta})$$

whenever $\xi,\eta \in H$.

Suppose that $\Phi_T \notin \overline{co}\,\{vTv^*: v \in G\}$. By the Hahn-Banach theorem, there exist a weakly continuous functional F on $\mathcal{L}(H)$ and $\alpha \in \mathbb{R}$, $\varepsilon \in \mathbb{R}_+^*$ such that

$$\mathit{Re}\ F(\Psi) \leq \alpha < \alpha + \varepsilon \leq \mathit{Re}\ F(\Phi_T)$$

whenever $\Psi \in \overline{co}\{vTv^* : v \in G\}$; we may choose $\xi_1,\ldots,\xi_n, \eta_1,\ldots,\eta_n \in H$ such that

$$F(\Psi) = \sum_{i=1}^{n} (\Psi\xi_i|\eta_i)$$

whenever $\Psi \in \mathcal{L}(H)$. In particular, we have

$$\mathit{Re} \sum_{i=1}^{n} (vTv^*\xi_i|\eta_i) = \mathit{Re}\ F(vTv^*) \leq \alpha$$

whenever $v \in G$. But, for every $v \in G$,

$$\sum_{i=1}^{n} f_{T,\xi_i,\eta_i}(v) = \sum_{i=1}^{n} (vTv^*\xi_i|\eta_i)$$

and, as M is a mean,

$$\mathcal{Re}\ F(\Phi_T) = \mathcal{Re} \sum_{i=1}^{n} (\Phi_T \xi_i | \eta_i)$$

$$\leq \sup \mathcal{Re}\ \{F(vTv^*) : v \in G\} \leq \alpha .$$

We come to a contradiction. Hence necessarily

$$\Phi_T \in \overline{\text{co}}\{vTv^* : v \in G\}.$$

On the other hand, for every $v_0 \in G$ and all $v \in G$, $\xi \in H$, $\eta \in H$,

$$_{v_0}f_{T,\xi,\eta}(v) = (v_0 v T v^* v_0^* \xi | \eta)$$

$$= (vTv^* v_0^* \xi | v_0^* \eta) = f_{T, v_0^*\xi, v_0^*\eta}(v).$$

As M is left invariant, for every $v_0 \in G$ and all ξ, $\eta \in H$, we have then

$$M(f_{T,\xi,\eta}) = M({}_{v_0}f_{T,\xi,\eta}) = M(f_{T,v_0^*\xi,v_0^*\eta}),$$

hence

$$(\Phi_T \xi | \eta) = (\Phi_T v_0^* \xi | v_0^* \eta) = (v_0 \Phi_T v_0^* \xi | \eta).$$

Thus, for every $v_0 \in G$, $v_0 \Phi_T v_0^* = \Phi_T$. We conclude that $\Phi_T \in A^{\sim}$. □

PROPOSITION 3.38: If A is a von Neumann algebra over the Hilbert space H admitting property (P), then there exists a projection $p \in \mathcal{L}(\mathcal{L}(H))$ of norm 1 applying $\mathcal{L}(H)$ onto A for which $p(\mathrm{id}_H) = \mathrm{id}_H$.

PROOF: If $T \in \mathcal{L}(H)$, we consider the closed convex hull A^T of $\{vTv^* : v \in A_u\}$ in the unit ball of $\mathcal{L}(H)$. Let S be the convex hull, closed in the $\sigma(\mathcal{L}(\mathcal{L}(H)), \mathcal{L}(H) \otimes_\gamma \mathcal{L}(H)_*)$ - topology, of the operators

$$T \mapsto vTv^*,$$
$$\mathcal{L}(H) \to \mathcal{L}(H)$$

where $v \in A_u$; it is contained in the unit ball of $\mathcal{L}(\mathcal{L}(H))$. Every $s \in S$ is positive and $s(id_H) = id_H$. We introduce a partial order in S putting $s' < s''$ for $s', s'' \in S$ in case $A^{s'(T)} \supset A^{s''(T)}$ for every $T \in \mathcal{L}(H)$. Let $(s_i)_{i\in I}$ be a net in S; for $j \in I$, we consider the closure S_j of $\{s_i \in S: s_j < s_i, i \in I\}$ in the above topology. As S is compact, we may choose $s \in \bigcap_{j\in I} S_j$. For every $i \in I$ and every $T \in \mathcal{L}(H)$, $A^{s_i(T)} \supset A^{s(T)}$ and s is an upper bound for $(s_i)_{i\in I}$. By Zorn's lemma there exists a maximal element p in S; p is positive, $p\,(id_H) = id_H$, $\|p\| = 1$.

We prove that, for every $T \in \mathcal{L}(H)$, $A^{p(T)}$ is a singleton. Assume that there exists $T_0 \in \mathcal{L}(H)$ such that $A^{p(T_0)}$ admits at least two elements. As $A^{p(T_0)} \cap A^{\sim} \neq \emptyset$, there exists $s_0 \in S$ such that $s_0(T_0) = T_1 \in A^{p(T_0)} \cap A^{\sim}$. Then $A^{T_1} = \{T_1\}$ and we have $A^{T_1} \subsetneq A^{p(T_0)}$, i.e., $A^{s_0(T_0)} \subsetneq A^{p(T_0)}$. We contradict the maximality of p. Hence, for every $T \in \mathcal{L}(H)$, $A^{p(T)}$ reduces to $\{p(T)\}$. Then

$$p^2(T) = p(p(T)) \in A^{p(T)} = \{p(T)\}.$$

Therefore p is a projection. □

<u>COROLLARY 3.39</u>: Every von Neumann algebra satisfying (P) is injective.

<u>PROOF</u>: The statement is an immediate consequence of Proposition 3.38. □

As a matter of fact, a von Neumann algebra is amenable if and only if it is approximately finite-dimensional, as well as, if and only if it satisfies (P). With respect to the equivalences established in part C and the preceding results, it suffices to show that every injective von Neumann algebra is approximately finite-dimensional. The proof is quite long; it relies on the decomposition of a von Neumann algebra into injective factors. We merely give some indications on the steps of the demonstration.

Let Γ be a Borel space equipped with a positive measure μ. One calls measurable field of Hilbert spaces on (Γ,μ) any family $\{H_\gamma : \gamma \in \Gamma\}$ of Hilbert spaces such that the following properties are satisfied with respect to a vector subspace E of $\prod_{\gamma\in\Gamma} H_\gamma$:

(1) For every $s = (s_\gamma) \in E$, the mapping $\gamma \mapsto \|s_\gamma\|_{H_\gamma}$, $\Gamma \to \mathbb{R}_+$ is μ-measurable.

(2) If $t = (t_\gamma) \in \prod_{\gamma\in\Gamma} H_\gamma$ and, for every $s = (s_\gamma) \in E$, the mapping $\gamma \mapsto (s_\gamma|t_\gamma)_{H_\gamma}$, $\Gamma \to \mathbb{C}$ is μ-measurable, then $t \in E$.

(3) There exists a sequence $(s^{(n)})$ in E such that, for every $\gamma \in \Gamma$, the family $\{s_\gamma^{(n)}\}$ is total in H_γ.

One now considers the family H of all $\xi \in E$ such that $\|\xi\| = (\int_\Gamma \|\xi_\gamma\|^2_{H_\gamma} d\mu(\gamma))^{\frac{1}{2}} < \infty$; H is a vector space. We may define a Hilbert structure on H via the sesquilinear form given by

$$(\xi|\eta) = \int_\Gamma (\xi_\gamma|\eta_\gamma)d\mu(\gamma)$$

for ξ, $\eta \in H$, where we identify ξ, $\eta \in H$ such that $\xi(\gamma) = \eta(\gamma)$ for μ-almost every $\gamma \in \Gamma$ ([T] p. 272). We say that H is a direct integral; it is denoted by

$$\int_\Gamma^\oplus H_\gamma \, d\mu(\gamma).$$

Every von Neumann algebra is a direct integral of factors ([T] Chapter IV, Proposition 8.21). A von Neumann algebra is injective if and only if almost all its factors are injective ([17] Proposition 6.5). Thus one has to establish that every injective factor of a von Neumann algebra over a separable Hilbert space is approximately finite-dimensional.

Let A be an injective factor of type II_1. For all $a_1,\ldots,a_n \in A_u$ and $\varepsilon > 0$, one may determine a finite subfactor B of A and $b_1,\ldots,b_n \in B$ such that $\|a_i-b_i\|_2 < \varepsilon$ whenever $i = 1,\ldots,n$.

Let A be an injective factor of type II_∞ and let p be a finite projection in A. Then, in A, A_p constitutes a factor of type II_1; there exists a separable Hilbert space H such that $A \simeq A_p \hat{\otimes} \mathcal{L}(H)$.

The class of factors of type III admits a classification into factors of types III_α $(0 \leq \alpha \leq 1)$. For every $\alpha \in \,]0,1[$, an injective factor of type III_α is the crossed product of a type II_∞ factor by a group generated by a single automorphism. An injective factor of type III_0 is a crossed product determined by an abelian von Neumann algebra. Moreover, every injective factor of type III_1 is also approximately finite-dimensional.

Finally we state amenability properties for a special kind of von Neumann algebras, the Kac algebras. They were considered mainly for the purpose of a general duality theory on locally compact groups. We first summarize the basic facts as given by Enock and J.-M. Schwartz [27].

Let A be a von Neumann algebra. Assume that there exist a) a normal injective *-morphism $\psi: A \to A \bar{\otimes} A$ such that $(\psi \otimes id_A) \circ \psi = (id_A \otimes \psi) \circ \psi$, b) an anti-*-automorphism ψ' : $A \to A$ such that $\psi' = \psi'^{-1}$ and $\iota \circ \psi \circ \psi' = (\psi' \otimes \psi') \circ \psi$, where ι is the automorphism of $A \bar{\otimes} A$ defined by $\iota(x \otimes y) = y \otimes x$ for $x, y \in A$. The triple (A, ψ, ψ') constitutes a *Hopf-von Neumann algebra*. For ω, $\omega' \in A_+$, we put

$$\langle x, \omega \square \omega' \rangle = \langle \psi(x), \omega \otimes \omega' \rangle,$$
$$\langle x, \omega^{\#} \rangle = \overline{\langle \psi'(x)^*, \omega \rangle},$$

$x \in A$. With respect to the multiplication $\square$ and the involution #, A_* constitutes an involutive Banach algebra.

Let now the Hopf-von Neumann algebra (A, ψ, ψ') admit also a weight ϕ (i.e., a positively homogeneous additive mapping of A_+ into $\overline{\mathbb{R}_+}$) that is normal, faithful, semifinite; let $t \mapsto \sigma_t^{\phi}$ be the canonically associated *-automorphism of A acting identically on the center of A ([65) p. 27). Then $K = (A, \psi, \psi', \phi)$ constitutes a *Kac algebra* if the following properties hold, u denoting the unit of A:

(i) For every $x \in A_+$, $(id_A \otimes \phi)(\psi(x)) = \phi(x)(u \otimes 1)$,

(ii) For all $x, y \in \{z \in A: \phi(z^*z) < +\infty\}$,

$$(id_A \otimes \phi)((u \otimes y^*)\psi(x)) = (\psi' \otimes \phi)(\psi(y^*)(u \otimes x)).$$

(iii) For every $t \in \mathbb{R}$, $\psi' \circ \sigma_t^{\phi} = \sigma_{-t}^{\phi} \circ \psi'$.

To the locally compact group G two important Kac algebras are associated.

1) $KA(G) = L^{\infty}(G \times G)$, identified with $L^{\infty}(G) \otimes L^{\infty}(G)$. For $f \in L^{\infty}(G)$,

$$\psi(f)(s,t) = f(st),$$

$$\psi'(f)(s) = f(s^{-1}),$$

$s,t \in G$. If $f \in L^{\infty}(G)_+$, let

$$\phi(f) = \int_G f d\lambda_G.$$

2) $KS(G) = PM(G) \subset \mathcal{L}(L^2(G))$. For $a \in G$,

$$\psi(L_a) = L_a \otimes L_a,$$

$$\psi'(L_a) = L_{a^{-1}}.$$

If $f \in C_+(G)$ and f admits a compact support, for the positive operator $T_f : g \mapsto f * g$ in $PM(G)$ one has $\phi(T_f) = f(e)$.
$L^2(G) \to L^2(G)$

Let H_ϕ be the Hilbert space determined by the trace ϕ of the Kac algebra $K = (A,\psi,\psi',\phi)$. One may introduce canonically, via the "Fourier representation" $\rho : A_* \to \mathcal{L}(H_\phi)$, the "dual" Kac algebra $\hat{K} = (\hat{A},\hat{\psi},\hat{\psi}',\hat{\phi})$. Then K is isomorphic to $\hat{\hat{K}}$. The elements of A_* [resp. $\hat{A}_*$] are given by $\omega_{\xi,\eta}(x) = (x\xi|\eta)$ [resp. $\hat{\omega}_{\xi,\eta}(x) = (x\xi|\eta)$] where $\xi, \eta \in H_\phi = H$, and $x \in A$ [resp. $x \in \hat{A}$]. Let J [resp. $\hat{J}$] be the involution defined by ϕ [resp. $\hat{\phi}$].

There exists a unitary operator $W = W_K \in (A \bar{\otimes} A)_u$ satisfying the following properties:

(i) $(\hat{J} \otimes J)W(\hat{J} \otimes J) = W^*$.

(ii) For every $x \in A$, $\psi(x) = W(u_A \otimes x)W^*$.

(iii) For all $\xi,\eta,\xi',\eta' \in H$, one has

$$(\hat{\rho}(\hat{\omega}_{\eta,\eta'})\xi|\xi') = (W(\xi\otimes\eta)|\xi'\otimes\eta') = (\eta|\rho(\omega_{\xi',\xi})\eta').$$

(iv) $W_{\hat{K}} = \sigma W_K^* \sigma$, with $\sigma(\xi \otimes \eta) = \eta \otimes \xi$ for $\xi, \eta \in H$. Also $\hat{\rho} = \psi' \circ {}^t\rho$ ([27] 4.3.9).

In particular, G being a locally compact group, one has $\widehat{KA(G)} = KS(G)$.

Let K be a Kac algebra defined for the von Neumann algebra A. The enveloping von Neumann algebra of A_* may be equipped with a Hopf-von Neumann algebra structure; its predual is denoted by B(K). If G is a locally compact group, B(KA(G)) is the Fourier-Stieltjes algebra B(G) of G ([P] p. 208); $B(KS(G)) = \mathcal{M}^1(G)$. Consider also the enveloping C*-algebra C*(K) of A_* and the C*-algebra $C^*_\rho(K)$ generated by ρ. The dual space $B_\rho(K)$ of $C^*_\rho(K)$ is a closed *-ideal of B(K); $B_\rho(K)$ is termed *Eymard algebra* of K.

PROPOSITION 3.40: Let $K = (A,\psi,\psi',\phi)$ be a Kac algebra over the Hilbert space H. The following properties are equivalent:

(i) There exists a net (ξ_i) of unitary vectors in H such that $\hat\rho(\hat\omega_{\xi_i})$ converges weakly to $\underset{\sim}{1} = id_H$.

(ii) There exists a net (ξ_j) of unitary vectors in H such that $\lim_i (\rho(\theta)\xi_i|\xi_i) = \theta(\underset{\sim}{1})$ whenever $\theta \in A_*$.

(iii) The algebra $B_\rho(K)$ is unital.

(iv) $B_\rho(K) = B(K)$.

(v) The canonical extension of ρ to C*(K) constitutes an isomorphism of C*(K) into $C^*_\rho(K)$.

(vi) There exists a net (ξ_j) of unitary vectors in H such that $\lim_i \|W(\eta \otimes \xi_i) - \eta \otimes \xi_i\| = 0$ whenever $\eta \in H$.

(vii) $\hat{A}_*$ admits bounded left approximate units.

PROOF:

(i) <=> (ii)

The condition (i) expresses the existence of a net (ξ_i) of unitary vectors in H such that, for every $\theta \in A_*$,

$$\lim_i \langle\hat\rho(\hat\omega_{\xi_i}), \theta \circ \psi'\rangle = \theta \circ \psi'(\underset{\sim}{1}) = \theta(\underset{\sim}{1}),$$

hence equivalently,

$$\lim_i \langle {}^t\rho(\hat{\omega}_{\xi_i}), \theta\rangle = \theta(\underset{\sim}{1}),$$

i.e.,

$$\lim_i (\rho(\theta)\xi_i | \xi_i) = \theta(\underset{\sim}{1}).$$

(ii) => (iii)

By hypothesis, for every $\theta \in A_*$,

$$|\langle \underset{\sim}{1}, \theta\rangle| \leqq \|\rho(\theta)\| .$$

If π denotes the homomorphism of A onto a von Neumann subalgebra of $\mathcal{L}(H_\phi)$, we have $\underset{\sim}{1} \in {}^t\pi(B_\rho(K))$. As ${}^t\pi$ is injective, $B_\rho(K)$ is unital.

(iii) => (iv)

As $B_\rho(K)$ is an ideal in B(K) and $B_\rho(K)$ is unital, $B_\rho(K) = B(K)$.

(iv) => (v) and (v) => (ii) are immediate implications.

(ii) => (vi)

By hypothesis, there exists a net (ξ_i) of unitary vectors in H such that, for any $\eta \in H$,

$$\lim_i (\rho(\omega_{\eta,\eta}) \xi_i | \xi_i) = \omega_{\eta,\eta}(\underset{\sim}{1}) = \|\eta\|^2;$$

thus also

$$\lim_i \|W(\eta \otimes \xi_i) - \eta \otimes \xi_i\|^2$$

$$= \lim_i (\|W(\eta \otimes \xi_i)\|^2 + \|\eta \otimes \xi_i\|^2 - 2\,\mathrm{Re}\ (W(\eta \otimes \xi_i) | \eta \otimes \xi_i))$$

$$= \lim_i 2(\|\eta \otimes \xi_i\|^2 - \mathrm{Re}\ (W(\eta \otimes \xi_i) | \eta \otimes \xi_i))$$

$$= \lim_i 2(\|\eta\|^2 - \mathrm{Re}\ (\rho(\omega_{\eta,\eta})\xi_i | \xi_i)) = 0.$$

(vi) => (vii)

Let $\eta \in H$. The net $(\xi_i)_{i \in I}$ being given, for every $i \in I$, we have

$$\| \hat{\omega}_{\xi_i} \,\square\, \hat{\omega}_\eta - \hat{\omega}_\eta \|$$

$$= \sup \{ | \langle \hat{\psi}(x), \hat{\omega}_{\xi_i} \otimes \hat{\omega}_\eta \rangle - \langle \hat{\psi}(x), \hat{\omega}_\eta \rangle | : x \in \hat{A}, \ \|x\| \leq 1 \}$$

$$= \sup \{ | (\sigma W^* \sigma(\underset{\sim}{1} \otimes x) \sigma W \sigma(\xi_i \otimes \eta) | \xi_i \otimes \eta) - (x\eta | \eta) | : x \in \hat{A}, \ \|x\| \leq 1 \}$$

$$= \sup \{ | ((x \otimes \underset{\sim}{1}) W(\eta \otimes \xi_i) | W(\eta \otimes \xi_i)) - ((x \otimes \underset{\sim}{1})(\eta \otimes \xi_i) | \eta \otimes \xi_i) | : x \in \hat{A}, \ \|x\| \leq 1 \}$$

$$\leq \sup \{ | \langle x \otimes \underset{\sim}{1}, \omega_{W(\eta \otimes \xi_i)} - \omega_{\eta \otimes \xi_i} \rangle | : x \in \hat{A}, \ \|x\| \leq 1 \}$$

$$\leq \| \omega_{W(\eta \otimes \xi_i)} - \omega_{\eta \otimes \xi_i} \|$$

$$= \| \tfrac{1}{2} (\omega_{W(\eta\otimes\xi_i) - \eta\otimes\xi_i, W(\eta\otimes\xi_i) + \eta\otimes\xi_i} + \omega_{W(\eta\otimes\xi_i) + \eta\otimes\xi_i, W(\eta\otimes\xi_i) - \eta\otimes\xi_i}) \|$$

$$\leq \|W(\eta \otimes \xi_i) + \eta \otimes \xi_i\| \ \|W(\eta \otimes \xi_i) - \eta \otimes \xi_i\|$$

$$\leq 2 \|\eta\| \ \|W(\eta \otimes \xi_i) - \eta \otimes \xi_i\| .$$

Thus by hypothesis

$$\lim_i \|\hat{\omega}_{\xi_i} \,\square\, \hat{\omega}_\eta - \hat{\omega}_\eta\| = 0;$$

then also

$$\lim_i \|\hat{\omega}_{\xi_i} \,\square\, \hat{\theta} - \hat{\theta}\| = 0$$

whenever $\theta \in \hat{A}_*$.

(vii) => (ii)

Let $(\hat{\theta}_i)$ be bounded left approximate units in $\hat{A}_*$; the net $(\hat{\rho}(\hat{\theta}_i))$ converges strongly to $\underset{\sim}{1}$. For every $\theta \in A_*$,

$$\lim_i \langle \hat{\theta}_i, \rho(\theta)\rangle = \lim_i \langle {}^t\rho(\hat{\theta}_i), \theta\rangle$$

$$= \lim_i \langle \hat{\rho}(\hat{\theta}_i), \theta \circ \psi'\rangle = \theta(\underset{\sim}{1}).$$

Then (ii) follows ([D'] 3.4.4). □

Let G be a locally compact group; G is amenable if and only if its Fourier algebra A(G), the predual of PM(G) = KS(G), admits bounded left approximate units ([P] Theorem 10.4).

Now one defines a *Kac algebra* to be *amenable* if it satisfies the equivalent conditions of Proposition 3.40. For a locally compact group G, $\widehat{KA(G)}_* = KS(G)_* = A(G)$; hence G is an amenable group if and only if KA(G) is an amenable Kac algebra.

If G is any locally compact group, the Kac algebra KS(G) is amenable. As a matter of fact, $\widehat{KS(G)}_* = L^1(G)$; as $L^1(G)$ admits bounded approximate units ([P] Proposition 2.2), property (vii) of Proposition 3.40 holds.

Notes

Bunce demonstrates 3.1 [8] and 3.2.3 [7].

Lau [54] studies in detail the unilateral amenability defined in 3.4; he establishes 3.5, the second half of the demonstration being modeled after Bonsall and Duncan [5]. Lau [54] proves 3.6.7 and several supplementary properties concerning the subject. He shows, for instance, that two Lau algebras A_1 and A_2 being given, $A_1 \oplus A_2$ is left amenable if and only if A_1 is; moreover, a Lau algebra A is left amenable if and only if A** is.

Semidiscrete von Neumann algebras are studied thoroughly by Effros and Lance [23]. They establish 3.9 - 3.18. In 1936 already, Murray and von Neumann [58] noticed that if A is a factor, the homomorphism ϖ, defined on $A \odot A^{\sim}$, is an isomorphism. Therefore the latter induces a norm on $A \odot A^{\sim}$; Effros and Lance [23] observe that by 3.18 this norm is an isometry if and only if the factor is semidiscrete. Lance [52] proves 3.19. See also Choi and Effros [13] [14].

Arveson [2] studied lifting problems for C*-algebras. Choi and Effros [14] considered the general phenomenon of injectivity in relation with operator spaces, terming the Banach space E injective provided any system of maps $\begin{matrix} A & \searrow & \\ \uparrow & & \\ B & \to & E \end{matrix}$ with respect to *-subspaces A, B of $\mathcal{L}(H)$ for a Hilbert space H, containing id_H, may be completed to a commutative diagram. Thus, in particular, the Hahn-Banach theorem states that the one-dimensional space $\tilde{R}$ is injective. Arveson [1] had generalized this theorem showing that $\mathcal{L}(H)$ is injective for any Hilbert space H. Choi and Effros [14] observed that an "operator system" in $\mathcal{L}(H)$ is injective if and only if there exists a completely positive projection of $\mathcal{L}(H)$ onto it.

Effros and Lance [23] give 3.21 - 3.23. The demonstrations of 3.24 - 3.27 are due to Haagerup [34]. Connes [18] establishes 3.28. Relying partially on Connes' work [17], Wassermann [70] proves 3.29 - 3.34.

De la Harpe [39] establishes 3.37. Property (P) is introduced by J.-T. Schwartz who proves 3.38 [62].

Hakeda and Tomiyama [38] studied the algebraic version of (P) as an extension property of the commutant of a von Neumann algebra. The von Neumann algebra A over the Hilbert space H is said to have the extension property if there exists a projection of norm 1 from $\mathcal{L}(H)$ onto A. They obtained an equivalent version that is independent of the underlying Hilbert space: For any C*-algebra B containing A there exists a projection of norm 1 from B onto A. In particular, if A admits (P), Schwartz's demonstration uses the extension property on $A^{\sim}$.

Kadison and Ringrose [48] noticed the amenability of any von Neumann algebra A admitting an amenable group of unitary operators the linear span of which is ultraweakly dense in A.

De le Harpe's result is exploited by Muhly and Saito [57] in their investigations on the analytic crossed product corresponding to an injective von Neumann algebra.

Effros and Lance [23] observe that the von Neumann algebra A is injective if and only if, for all C*-algebras B, B_1 such that $B \subset B_1$, one has $A \otimes_{nor} B \subset A \otimes_{nor} B_1$. Let G be a discrete group and consider a covariant system (A, G, α) for the C*-algebra A; denote by α' the induced action of G on the enveloping von Neumann algebra A**. Assume the existence of a central projection $p \in A^{**}$ such that $\sum_{x \in G} \alpha'_x(p) = 1$ and $\alpha'_x(p)p = 0$ whenever $x \in G \setminus \{e\}$. Chu [16] considers an associated C*-algebra crossed product and proves it to be nuclear if and only if A is.

If G is an amenable group, then the C*-algebra of G, generated by all unitary representations of G and denoted by C*(G), coincides with $C_L^*(G)$ ([P] Theorem 8.9). Guichardet [31] shows that in this case C*(G) is nuclear; hence it is amenable. Green [30] obtains a general version for the homogeneous space G/H of a locally compact group G by a closed subgroup H, that is amenable, i.e., $C(G/H)$ admits a mean, invariant for the action of G on G/H ([P] p. 364).

Bunce and Paschke [9] call quasi-expectation of a unital C*-algebra A onto a unital C*-subalgebra B any projection $p \in \mathcal{L}(A,B)$, such that $p(sxt) = sp(x)t$ whenever $x \in A$, $s, t \in B$. They show that if, for the von Neumann algebra A over the Hilbert space H, there exists a quasi-expectation of $\mathcal{L}(H)$ onto A, then A is injective, and that if B is an amenable von Neumann subalgebra of the von Neumann algebra A, then there exists a quasi-expectation of A onto $B^{w} \cap A$. As corollaries they state that any amenable C*-algebra is nuclear and any amenable von Neumann algebra is injective. Bunce and Paschke [10] establish a direct demonstration showing that if A is a nuclear C*-algebra and X is a Banach A-module with weakly sequentially complete dual space X*, then every continuous derivation D of A into X is of the form D(a) = af-fa (a ∈ A) for an element f in X**. Haagerup [34] proves that for an infinite-dimensional Hilbert space H and a nuclear C*-algebra A, the C*-algebra $A \hat{\otimes} \mathcal{L}C(H)$ is strongly amenable.

Approximately finite-dimensional von Neumann algebras are studied originally under the term hyperfinite von Neumann algebras. Elliott [24] furnishes a first classification of the biduals of separable approximately finite-dimensional C*-algebras; see also Elliott [25], Elliott and Woods [26]. Connes [17] proves that a von Neumann algebra is injective if and only if almost all of its factors are injective. Connes realizes the remarkable classification of the injective factors of the different types and shows that every injective factor is approximately finite-dimensional [17] [15] [16]. The theory is exposed by Wright [72]. See also Takesaki [67]. Haagerup [36] elaborates new proofs of the fact that every injective von Neumann algebra factor of type II_1 or of type II_∞ is approximately finite-dimensional. Relying on Connes' work, Choi and Effros demonstrate that if A is nuclear C*-algebra, its enveloping von Neumann algebra A** is injective in case A is separable [11], in the general case [12].

A concise, instructive exposition on the content and the importance of the theory of approximately finite-dimensional von Neumann algebras is Moore [55]. See also Strătilă [65].

The theory of amenable algebras is explited by Connes, Feldman and Weiss in their investigations on equivalence relations possessing amenability properties [21]. The subject is studied by Moore [56].

To a certain extent amenability on von Neumann algebras is a replica of amenability on discrete groups. Among the possible illustrations we merely select the canonical correspondence: The role played on the amenable discrete group G by an invariant mean for $\ell^{\infty}(G)$ is taken, on a von Neumann algebra A over a Hilbert space H, by a hypertrace, i.e., a state M on $\mathcal{L}(H)$ such that $M(aT) = M(T) = M(Ta)$ whenever $T \in \mathcal{L}(H)$ and $a \in A_u$.

The class of Kac algebras was introduced in order to obtain duality properties on arbitrary locally compact groups by Vainermann and Kac [68], as well as by Enock and J.-M. Schwartz [27], J.-M. Schwartz [63]. It was studied extensively by Strătilă [65].

Voiculescu [70] established first amenability properties for Kac algebras. The proofs of 3.40 are due to Enock and J.-M. Schwartz; they are summarized in [28], laid out in [29]. In analogy to amenability for locally compact groups, these authors obtain a collection of supplementary characterizations of amenability, namely fixed point properties, for Kac algebras.

Basic references

[B] Bourbaki, Nicolas. Eléments de mathématique. Espaces vectoriels topologiques. Chapters 1-5, Masson, Paris, 1981.

[D] Dixmier, Jacques. Les algèbres d'opérateurs dans l'espace hilbertien. Second edition. Gauthier-Villars, Paris, 1969.

[D'] Dixmier, Jacques. Les C*-algebres et leurs représentations. Second edition. Gauthier-Villars, Paris, 1969.

[DS] Dunford, Nelson, and Jacob T. Schwartz. Linear Operators. Part I. General Theory. Third edition. Interscience, New York, 1966.

[P] Pier, Jean-Paul. Amenable Locally Compact Groups. John Wiley, New York, 1984.

[S] Sakai, Shôichirô. C*-Algebras and W*-Algebras. Springer, Berlin, 1971.

[T] Takesaki, Masamichi. Theory of Operator Algebras I. Springer, Berlin, 1979.

Bibliography

[1] Arveson, William B. Subalgebras of C*-algebras I. Acta Math. 123, 141-224 (1969); 128, 271-308 (1972).

[2] Arveson, William B. Notes on extensions of C*-algebras. Duke Math. J. 44, 329-355 (1977).

[3] Barnes, Bruce A. When is a representation of a Banach *-algebra Naimark-related to a *-representation? Pacific J. Math. 72, 5-25 (1977).

[4] Blackader, Bruce E. Weak expectations and nuclear C*-algebras. Indiana Univ. Math. J. 27, 1021-1026 (1978).

[5] Bonsall, F.F., and J. Duncan. Complete normed algebras. Springer, Berlin, 1973.

[6] Bunce, John W. Characterizations of amenable and strongly amenable C*-algebras. Pacific J. Math. 43, 563-572 (1972).

[7] Bunce, John W. Representations of strongly amenable C*-algebras. Proc. Amer. Math. Soc. 32, 241-246 (1972).

[8] Bunce, John W. Finite operators and amenable C*-algebras. Proc. Amer. Math. Soc. 56, 145-151 (1976).

[9] Bunce, John W., and William L. Paschke. Quasi-expectations and amenable von Neumann algebras. Proc. Amer. Math. Soc. 71, 232-236 (1978).

[10] Bunce, John W., and William L. Paschke. Derivations on a C*-algebra and its double dual. J. Functional Analysis 37, 235-247 (1980).

[11] Choi, Man-Duen, and Edward G. Effros. Separable nuclear C*-algebras and injectivity. Duke Math. J. 43, 309-322 (1976).

[12] Choi, Man-Duen, and Edward G. Effros. Nuclear C*-algebras and injectivity: The general case. Indiana Univ. Math. J. 26, 443-446 (1977).

[13] Choi, Man-Duen, and Edward G. Effros. Lifting problems and the cohomology of C*-algebras. Canad. J. Math. XXIX, 1092-1111 (1977).

[14] Choi, Man-Duen, and Edward G. Effros. Injectivity and operator spaces. J. Functional Analysis 24, 156-209 (1977).

[15] Choi, Man-Duen, and Edward G. Effros. Nuclear C*-algebras and the approximation property. Amer. J. Math. 100, 61-79 (1978).

[16] Chu, Cho-Ho. A note on crossed products of nuclear C*-algebras. Rev. Roumaine Math. Pures Appl. XXX, 99-102 (1985).

[17] Connes, Alain. Classification of injective factors. Ann. of Math. 104, 73-115 (1976).

[18] Connes, Alain. On the cohomology of operator algebras. J. Functional Analysis 28, 248-253 (1978).

[19] Connes, Alain. Von Neumann algebras. Proceedings of the International Congress of Mathematicians, Helsinki, 1978, vol. I, 97-109. University of Helsinki, 1980.

[20] Connes, Alain. Classification des facteurs. Proceedings of Symposia in Pure Mathematics, vol. 32/2, 43-109. American Mathematical Society, Providence, 1982.

[21] Connes, Alain, Jacob Feldman, and Benjamin Weiss. An amenable equivalence relation is generated by a single transformation. Ergod. Th. and Dynam. Sys. 1, 431-450 (1981).

[22] Cuntz, Joachim. Simple C*-algebras generated by isometries. Commun. Math. Phys. 57, 173-185 (1977).

[23] Effros, Edward G., and E. Christopher Lance. Tensor products of operator algebras. Advances in Math. 25, 1-34 (1977).

[24] Elliott, George A. On approximately finite-dimensional von Neumann algebras. Math. Scand. 39, 91-101 (1976).

[25] Elliott, George A. On approximately finite-dimensional von Neumann algebras II. Canad. Math. Bull. XXI, 415-418 (1978).

[26] Elliott, George A. and E.J. Woods. The equivalence of various definitions for a properly infinite von Neumann algebra to be approximately finite-dimensional. Proc. Amer. Math. Soc. 60, 175-178 (1976).

[27] Enock, Michel, and Jean-Marie Schwartz. Une dualité dans les algèbres de von Neumann. Mém. Soc. Math. France 44, 1-144 (1975).

[28] Enock, Michel, and Jean-Marie Schwartz. Moyennabilité des groupes localement compacts et algèbres de Kac. C.R. Acad. Sci. Paris 300, 625-626 (1985).

[29] Enock, Michel, and Jean-Marie Schwartz. Algèbres de Kac moyennables. Pacific J. Math. 125, 363-379 (1986).

[30] Green, Philip. The local structure of twisted covariance algebras. Acta. Math. 140, 191-250 (1978).

[31] Guichardet, Alain. Tensor products of C*-algebras. Aarhus Universitet Lecture Notes Series 12, 1969.

[32] Guichardet, Alain. Systèmes dynamiques non commutatifs. Astérisque 13-14, 1974.

[33] Guichardet, Alain. Cohomologie des groupes topologiques et des algèbres de Lie. Cedic/Fernand Nathan, Paris, 1980.

[34] Haagerup, Uffe. All nuclear C*-algebras are amenable. Invent. Math. 74, 305-319 (1983).

[35] Haagerup, Uffe. Solution of the similarity problem for cyclic representations of C*-algebras. Ann. of Math. 118, 215-240 (1983).

[36] Haagerup, Uffe. A new proof of the equivalence of injectivity and hyperfiniteness for factors on a separable Hilbert space. J. Functional Analysis 62, 160-201 (1985).

[37] Haagerup, Uffe. The Grothendieck inequality for bilinear forms on C*-algebras. Advances in Math. 56, 93-116 (1985).

[38] Hakeda, Jôsuke, and Jun Tomiyama. On some extension properties of von Neumann algebras. Tôhoku Math. J. 19, 315-323 (1967).

[39] Harpe, Pierre de la. Moyennabilité du groupe unitaire et propriété P de Schwartz des algèbres de von Neumann. Proceedings: Algèbres d'opérateurs. Lecture Notes in Mathematics, vol. 725, 220-227. Springer, Berlin, 1979.

[40] Hewitt, Edwin, and Kenneth A. Ross. Abstract harmonic analysis; vol. I, second edition, 1979; vol. II, 1970. Springer, Berlin.

[41] Hewitt, Edwin, and Karl Stromberg. Real and abstract analysis. Second edition. Springer, Berlin, 1969.

[42] Johnson, Barry E. Cohomology in Banach algebras. Mem. Amer. Math. Soc. 127 (1972).

[43] Johnson, Barry E. Approximate diagonals and cohomology of certain annihilator Banach algebras. Amer. J. Math. 94, 685-698 (1972).

[44] Johnson, Barry E. Introduction to cohomology in Banach algebras. Algebras in Analysis, 84-100. Academic, New York, 1975.

[45] Johnson, Barry E. Perturbation of Banach algebras. Proc. London Math. Soc. 34, 439-458 (1977).

[46] Johnson, Barry E., Richard V. Kadison, and John R. Ringrose. Cohomology of operator algebras, III. Reduction to normal cohomology. Bull. Soc. Math. France 100, 73-96 (1972).

[47] Kadison, Richard V. The von Neumann algebra characterization theorems. Expo. Math. 3, 193-227 (1985).

[48] Kadison, Richard V., and John R. Ringrose. Fundamentals of the theory of operator algebras, vol. II. Academic Press, Orlando, 1986.

[49] Kaijser, Sten, and Allan M. Sinclair. Projective tensor products of C*-algebras. Math. Scand. 55, 161-187 (1984).

[50] Khelemskiĭ, A. Ya., and M.V. Sheinberg. Amenable Banach algebras. Funkcional Anal. i. Priložen 13, 42-48 (1979). English translation: Functional Anal. Appl. 13, 32-37 (1979).

[51] Kirchberg, Eberhard. C*-nuclearity implies CPAP. Math. Nachr. 76, 203-212 (1977).

[52] Lance, E. Christopher. On nuclear C*-algebras. J. Functional Analysis 12, 157-176 (1973).

[53] Lau, Anthony To-Ming. Characterizations of amenable Banach algebras. Proc. Amer. Math. Soc. 70, 156-160 (1978).

[54] Lau, Anthony To-Ming. Analysis on a class of Banach algebras with applications to harmonic analysis on locally compact groups and semigroups. Fund. Math. 118, 161-175 (1983).

[55] Moore, Calvin C. Approximately finite von Neumann algebras. Amer. Math. Monthly 85, 657-659 (1978).

[56] Moore, Calvin C. Ergodic theory and von Neumann algebras. Proceedings of Symposia in Pure Mathematics, vol. 38/2, 179-226. American Mathematical Society, Providence, 1982.

[57] Muhly, Paul S., and Kiche-Suke Saito. Analytic crossed products and outer conjugacy classes of automorphisms of von Neumann algebras. Math. Scand. 58, 55-68 (1986).

[58] Murray, F.J., and John von Neumann. On rings of operators. Ann. of Math. 37, 116-229 (1936).

[59] Paschke, William L. The crossed product of a C*-algebra by an endomorphism. Proc. Amer. Math. Soc. 80, 113-118 (1980).

[60] Racher, Gerhard. On amenable and compact groups. Monats. Math. 92, 305-311 (1981).

[61] Rosenberg, Jonathan. Amenability of crossed products of C*-algebras. Comm. Math. Phys. 57, 187-191 (1977).

[62] Schwartz, Jacob T. Two finite, non-hyperfinite, non-isomorphic factors. Comm. Pure Appl. Math. 16, 19-26 (1963).

[63] Schwartz, Jean-Marie. Sur la structene des algèbres de Kac, I. J. Funct. Anal. 34, 370-406 (1979).

[64] Stegmeir, Ulrich. Approximierende Einsen in Idealen von Gruppenalgebren. Dissertation, Technische Univ. Munich, 1978.

[65] Strǎtilǎ, Serban. Modular Theory in Operator Algebras. Editura Academiei, Bucarest; Abacus Press, Tunbridge Wells, U.K., 1981.

[66] Takesaki, Masamichi. On the crossnorm of the direct product of C*-algebras. Tôkoku Math. J. 16, 111-112 (1964).

[67] Takesaki, Masamichi. Structure of factors and automorphism groups. American Mathematical Society, Providence, 1983.

[68] Vainermann, L.I., and G.I. Kac. Non-unimodular ring groups and Hopf von Neumann algebras. Mat.Sb. 23, 185-214 (1974). English translation: Math. USSR-Sb. 94, 194-225 (1974).

[69] Van Daele, Alfons. Continuous crossed products and type III von Neumann algebras. Lecture Notes Series 31. London Mathematical Society. Cambridge University Press, 1978.

[70] Voiculescu, Dan. Amenability and Katz algebras. Algèbres d'opérateurs et leurs applications en physique mathématique. 274. Editions CNRS, Paris, 1979.

[71] Wassermann, Simon. Injective W*-algebras. Math. Proc. Camb. Phil. Soc. 82, 39-47 (1977).

[72] Wright, Steve. Classification of injective factors: The work of Alain Connes. Intern. J. Math. & Math. Sci. 6, 1-39 (1983).

[73] Zillmann, Hartmut. Tensorprodukte von C*-Algebren und vollständig positive Operatoren. Diss. Eberhard-Karls-Univ. Tübingen, 1983.

Index of notations

Index of terms